ISBN 978-1-333-73349-0
PIBN 10540784

1 MONTH OF FREE READING

at

www.ForgottenBooks.com

By purchasing this book you are eligible for one month membership to ForgottenBooks.com, giving you unlimited access to our entire collection of over 700,000 titles via our web site and mobile apps.

To claim your free month visit:

www.forgottenbooks.com/free540784

English
Français
Deutsche
Italiano
Español
Português

www.forgottenbooks.com

Mythology Photography **Fiction**
Fishing Christianity **Art** Cooking
Essays Buddhism Freemasonry
Medicine **Biology** Music **Ancient
Egypt** Evolution Carpentry Physics
Dance Geology **Mathematics** Fitness
Shakespeare **Folklore** Yoga Marketing
Confidence Immortality Biographies
Poetry **Psychology** Witchcraft
Electronics Chemistry History **Law**
Accounting **Philosophy** Anthropology
Alchemy Drama Quantum Mechanics
Atheism Sexual Health **Ancient History**
Entrepreneurship Languages Sport
Paleontology Needlework Islam
Metaphysics Investment Archaeology
Parenting Statistics Criminology
Motivational

The Scottish Botanical Review

A QUARTERLY MAGAZINE

(Including the Transactions of the Botanical Society of Edinburgh)

Edited by M'Taggart Cowan, Jr.

EDITORIAL COMMITTEE

William Barclay. Robert H. Meldrum.
Arthur Bennett, A.L.S. W. G. Smith, B.Sc., Ph.D.
A. W. Borthwick, D.Sc. James Stirton, M.D., F.L.S.

EDINBURGH
NEILL & CO. LTD.
Bellevue

Price 2s. 6d.   Annual Subscription, payable in advance, 7s. 6d., post free in U.K.

SUBSCRIPTIONS

Subscriptions are payable in advance by cheque or postal order to M'TAGGART COWAN, Jr., 40 Great King Street, Edinburgh.

Single copies from Messrs. Neill & Co., Edinburgh.

CONTRIBUTIONS

Contributions on all subjects connected with botanical science and teaching will be fully considered. A preference will be given to contributions relating particularly to Scotland.

All manuscript must be clearly written, and when possible typed.

Illustrations should if possible be adapted for reproduction in the text—that is, pen-and-ink drawings or photographs. Where special plates are desired, contributors should first communicate with the Editor, as only a limited number of these can be published.

Authors will receive 25 copies of their communications free. Additional copies, in covers, may be had from the Printers at the undermentioned rates, provided such orders accompany the manuscript.

	25 copies	50 copies	100 copies.
8 pages .	3s. 6d.	5s. 0d.	8s. 0d.
16 ,, . .	5s. 6d.	8s. 0d.	13s. 0d.
32 ,, . .	9s. 6d.	13s. 0d.	20s. 6d.

All communications on Editorial matters, Advertisements, and Books for Notice should be addressed to M'TAGGART COWAN, Jr., 40 Great King Street, Edinburgh.

The
Scottish Botanical Review

No. 1]   1912   [JANUARY

The Geological Relations of Stable and Migratory
Plant Formations. By C. B. Crampton, M.B.,
C.M., of H.M. Geological Survey.

PART I. INTRODUCTION.

THE nature of a plant formation is still largely a matter for
investigation, even though certain groups of plant associations,
for instance those of the moorland, are usually conceded as
plant formations.   On the other hand, it is pretty generally
admitted that it is the habitat that determines plant association in at least its more comprehensive units ; but the factors
which contribute to the habitat in such cases are so complex
and interwoven that plant formations have often been defined
by limited groups of factors considered in conjunction to
be the master-factors under existing conditions.   Thus
Schimper (1) emphasised the climate as a master-factor in
formations he distinguished as climatic formations from
others he termed edaphic formations, where he considered
the soil factors of greater importance ; and Graebner has
proposed a classification based on the richness or poverty of
the soil waters in mineral matters (2).   The soil, again, is

adjudged the master-factor in plant formation in the " Types of British Vegetation" recently published (3). Many still, perhaps advisedly, refer to forest and meadow formations, thus falling back on a physiognomic nomenclature in default of finding a master-factor other than the sum total of the complex relations of these plant associations.

Starting from another standpoint, and following Warming, Cowles and Clements emphasised the succession of plant invasions on the appearance of a new habitat. Cowles pointed out that successions may be progressive or retrogressive in nature (4). Clements applied the term "formation" to certain stages of plant association encountered in the successions. He also framed a number of laws of succession, of which the following appear to be the most important in respect of the present discussion :—

I. "The initial cause of a succession is the formation or appearance of a new habitat, or a striking change in an existing one." It should be remarked here that a change in an existing habitat is probably of equal or greater importance than the formation of a new one.

2. "Each stage of a succession reacts upon the habitat in such a way as to produce conditions more or less unfavourable to itself, but favourable to the invaders of the next stage."

3. "Initial formations are open, ultimate formations are closed." To this might be added, except under semi-desert conditions.

4. "The universal tendency of vegetation is towards stabilisation."

5. "The ultimate stage of a succession is determined by the dominant vegetation of the region. Lichen formations are often final in polar and niveal zones. Grassland is the final vegetation for plains and alpine stretches, and for much prairie, while forest is the last stage for all mesophytic habitats."

6. "The end of a succession is largely brought about by the progressive increase and competition, which makes the entrance of invaders more and more difficult." This, of course, only embraces closed plant formations (5).

Moss has recently laid particular stress on the succession of associations within the same formation, and defines a

plant formation as follows :—" A plant formation comprises
the progressive associations which culminate in one or more
stable or chief associations, and the retrogressive associations
which result from the decay of the chief associations, so long
as these changes occur on the same habitat " (6).

The crucial point in this definition is the meaning of the
term "the same habitat," since in most cases of known
progression of associations the habitat suffers change during
the various stages of succession.   The meaning of "the same
habitat " in this definition is made somewhat clearer in the
" Types of British Vegetation," where the master-factor involved in plant formation is considered to be the nature of
the soil.   In the case of certain formations therein described
(as, *e.g.*, 1, the plant formation of clays and loams ; 2,
the plant formation of sandy soil; 3, the plant formation
of older siliceous soils ; 4, the plant formation of calcareous soils) a wide margin of soil difference is allowed
within the meaning of "the same habitat" which defines the
plant formation, since the very different degrees of water
content, aeration, humus content, warmth, etc., due to
drainage, exposure, and depth of soil, *i.e.* the physiography
of these formations, has as yet received but little attention.

Clays, loams, and sands, generally speaking, form level or
gently undulating ground on valley floors, plains, or plateaux.
In humid regions not recently glaciated all extensive surfaces
of this nature are cloaked by these deposits, irrespective of
the nature of the underlying rocks.   This is due to clays and
sands, pure or in varying mixtures, forming the ultimate products of decomposition of all rocks where the surface has
long been subject to decay without erosion, and further
forming the products of accumulation in areas subject to
sedimentation.   Wherever the ground slopes, clays and sands
are removed by denudation.

In large continental areas they cloak plains and plateaux
at very different altitudes, with a wide range in climatic
conditions, and correspondingly wide changes in the vegetation.   In the British Isles, as in all smaller areas of fairly
mature physiography, such materials are, however, chiefly
confined to lower levels, and the plant formations cover-
ng them might well be termed collectively the plant forma-

tions of the lower ground.   Where such soils form the lower ground in Britain, they have all the advantages of shelter from the wind and of summer warmth, and they also lie in the direct route of migration from the Continent, factors favouring the presence of the later post-glacial immigrants and a forest vegetation.   Clays and sands in the northern and western parts of our islands, and those at high levels, have suffered encroachment by moorland almost as much as any other type of soil ; but sands and clays everywhere differ in the former being more subject to leaching and consequent encroachment by heath, while the clays tend rather to stagnation of soil waters in low-lying positions.

On the other hand, the so-called siliceous soils derived from the harder sandstones, grits, and metamorphic rocks usually form sloping, hilly ground, and in this country are therefore more subject to atmospheric humidity, cool summer temperatures, and leaching.   This is due to the past relations of the rocks to denudation, *i.e.* to physiography.

The relations of the calcareous soils to physiography and vegetation are very complex, and will be referred to later. In the " Types of British Vegetation " the retrogressive changes due to human and other interference in the woodlands are discussed in detail, but we find no account of progressive association in this connection, and have to turn to other formations for their demonstration.   Analysis of any published list of plant formations shows, indeed, that the progressive associations commonly met with are those belonging to the types of successions to which Cowles has given the names " topographic " and " biotic," while the retrogressive associations are chiefly distinctive of successions which have at least partly completed stabilisation, and whose initial stages are in most cases unknown.

Cowles has recently defined three cycles of vegetative succession :—

1. *Regional successions*, " due to secular change, and in rate of development bearing some comparison with the succession of geological periods."   Of these the most important to the plant ecologist in this country is " the post-glacial invasion of southern forms into boreal regions, accompanying and following the retreat of the ice " (7).

2. *Topographic successions*, "of much greater rapidity, and associated with the topographical changes resulting from the activities of such agents as running water, wind, ice, gravity, and vulcanism, and leading in general to erosion and deposition." In this respect he points out that "the influence of erosion is generally destructive to vegetation, or at least retrogressive, *i.e.* tending to cause departure from the mesophytic, while the influence of deposition is constructive or progressive, *i.e.* tending to cause an approach towards the mesophytic."

3. *Biotic successions*, where the vegetative changes are due to plant and animal agencies. "If in their operation regional agencies are matters of æons, and topographical agencies matters of centuries, biotic agencies may be expressed in terms of decades." "The influence of biotic agencies is not confined to areas that are characterised by a *pre-erosion topography*, because the interval between the periods of active erosion often is sufficiently long to permit the development of an entire biotic cycle."

This is a great step forward in plant ecology, since it is the first indication of an appreciation of the difference in origin between plant communities covering wide areas of the earth's crust, which are, geologically speaking, in a stable condition for the time being, and others taking part in the topographic successions with which we are now to some extent familiar.

A point of first importance in defining the habitat of a plant association is its present and late condition relative to the geological agents of surface change. Is the relation between the habitat and the vegetation comparatively stable, and inherited from the past as such? or is the vegetation in various stages of progressive association owing to constantly recurring changes in the geology of the habitat, and can stages in the progress of associations be demonstrated?

We would therefore distinguish two classes of habitats which differ in plant succession and in the limits set to stabilisation. We would further distinguish two classes of plant formations which differ in their centres of distribution, but which tend to overlap and invade one another's territory owing to the migratory nature of the geological agents of surface change. The *stable formations* are those whose plant associations have their centres of distribution on ground which has been for a long period comparatively stable from the geological standpoint, and under climatic conditions favourable to the type of vegetation. The *migratory formations* are those whose plant associations have their centres of distribution in areas within the sphere of influence of the geological agents of surface change, and only naturally initiate and progress on habitats directly generated and periodically or continuously influenced by such agents.

The factors defining these two classes of habitats are as follows:—

In the *stable formations* they are: 1, the climate, as due to geography, the amount and seasonable combinations of light, heat, wind, atmospheric moisture, and rain ; 2, the nature of the soil, resulting from the geological distribution of the rocks in their present and past relations to climate, vegetation, and geographical change.

In the *migratory formations*: 1, the relation of habitat to climate is chiefly topographical; and 2, the soil varies in each case with the nature of the geological agent of surface change, its topographical relations to erosion or deposition, its constant, periodical, or occasional character, and the stage and kind of succession of plant association growing upon it.

The plants of the dominant stable formations are constantly attempting to colonise the habitats of the adjacent migratory formations, but only succeed in so doing after the geological agents have ceased for some time to exert their influence. On a rock cliff, for instance, species which are normal to the habitat, as lithophytes, or as chomophytes (either endemic species or those incapable of withstanding competition), are found side by side with invading species from the neighbouring stable formations, but the latter can never succeed in displacing the former species until the habitat is permanently altered in its relations to the geological agents of surface change. In deserted quarries, at a distance from naturally formed rock formations, invasion is rapid, and the species other than lithophytes may entirely consist of such invaders. Closed migratory formations, as the reed-belt, marsh, fen, and alluvial meadow, are generally free from invaders until degeneration sets in.

On the other hand, interference by man or beast with the habitat of a stable formation often permits of a temporary invasion by plants which are normally excluded from it, and which have their natural station on migratory habitats. Stable formations that are undergoing retrogression through climatic change may also show partial invasion by plants from the neighbouring migratory formations. The present invasion of the moorland by extensive colonies of *Nardus stricta* is probably to be looked upon as of this nature. *Nardus* is normally confined in moorland districts to the flood-rims of burns and " flushes," where the drainage comes from acid peat or peaty hollows subject to snow-lie.

From this tendency on the part of the plants of the stable formations to eventually obtain possession of areas formerly occupied by migratory formations on the one hand, and of certain plants of the migratory formations to invade disturbed areas of the stable formations on the other, a feeling appears to have arisen that the succession of plant associations is always from those characteristic of the neighbouring migratory formations towards the full stabilisation met with in the stable formations. But most of the plant associations of the migratory formations, such as those of rock, marsh, aquatic, and saline habitats, are bound down to sharply defined physical environments limited to the immediate neighbourhood of the migratory agents of change ; and, with regard to many of the plants of the migratory formations, it is doubtful if at the initiation of the existing stable formations they had yet migrated to their present limits of distribution. It is, moreover, fairly certain that with change of climate stable types of vegetation may gradually replace other stable types without complete initiations of succession such as are met with in the migratory habitats. Where large regions devoid of vegetation arise rapidly *de novo*, as in volcanic areas, the whole surface consists of bare rock, and the succession of plant associations must initiate as lithophytes. These cases are, however, exceptional, since large regions of the globe are known to have existed chiefly under a deep load of soils and subsoils for such long periods that they must have undergone many vicissitudes of climate and changes of plant formations.

The plant associations of stable formations are stabilised to the extent allowed by the prevailing climate and edaphic conditions, and are of long persistence on the same habitat, but their boundaries are subject to frequent retraction and expansion owing to the migratory nature of the habitats of the migratory formations. Moreover, under a continuance of the same climatic and geographic conditions, their closed plant associations appear capable of resisting invasion. In open stable formations the physical conditions limit invasion. Change in stable formations is therefore in all probability to be chiefly accounted for by change of climate or geography, inducing accommodative changes in the plant associations or dislocation in the balance of relations between the associations and their habitats and those of neighbouring plant formations.

Migratory formations are of comparatively short persistence on the same habitat, which sooner or later undergoes change or destruction, with renewal elsewhere. On retreat of the geological agent, their associations tend to rapid degeneration from plant invasion. All stages of progressive successions of associations are encountered, owing to the frequent formation of new habitats and the zonal nature of the influence of the geological agents.

PART II. THE REGIONAL SUCCESSION OF STABLE FORMATIONS.

The replacement of one type of vegetation by another over wide areas of the earth's crust has doubtless always been partly due to the secular migration and competition of plants and animals, but probably chiefly originates in extensive geological transformations or secular changes in climate and geography. Such changes may occur :—

1. From tangential movements along lines of weakness in the earth's crust, leading to mountain-building, and thus erecting barriers to moisture-laden winds, and new belts or foci involving climatic zonation due to altitude. The more deeply seated rocks implicated in these tangential movements usually undergo more or less metamorphism and injection by igneous magma, and cores of ancient crystalline rocks are

drawn up in the folds ; and crystalline gneisses and plutonic rocks exposed by denudation of the mountain masses, being generally more resistant to erosion than unmetamorphosed sediments, contribute to their permanence. These movements also have a special tendency to recur at intervals over long periods of time, and the resulting mountains form major orographical features of the surface of the globe. On cessation of the movements, the mountains are gradually removed by denudation, and wide plateaux and plains of crystalline gneisses may form the sole geological evidence of their former existence. From time to time in geological history, new lines of weakness have developed with a new orientation, causing profound changes in geography, affecting the ocean currents, the climate, the distribution of rocks and their relations to erosion.

2. From upward or downward displacements of the strand-line, either limited in extent and due to orogenic or to isostatic movements, or world-wide in their effects. The latter, according to Suess, are due to accumulation of sediment on the sea-bottom, or subsidence of the ocean floor (8), and Penck has suggested that the formation of great ice-caps may affect the sea level the world over (9). Such changes of level lead to diminution or expansion of continents, and alterations in their relations to islands and the areas subject to erosion, with changes in the barriers to plant migration, in the distribution of ocean currents, and in the range of humid, maritime, arid, and continental climates. Suess has suggested practically world-wide· transgressions of the sea over the land in the Middle Devonian, Middle and Late Carboniferous, Early and Middle Jurassic, and Cretaceous times. On the other hand, during the Lower Devonian, Late Devonian, and Early Carboniferous, the New Red Sandstone and Late Jurassic, there were negative phases or downward displacements of the strand-line. There is good reason for believing that widely distributed geological deposits belonging to the negative phases are of continental and frequently of desert origin, and the known fossil fauna and the nature of the associated rocks point to life in and around inland seas and lakes. The continental areas of desert and small rainfall may have had a different relation

to the tropics, or a peculiarly wide distribution during such phases. Suess believes that the negative phases were more or less spasmodic in character, such as might lay bare in short periods, geologically speaking, wide areas of the marine platforms of the continental shelves. In certain cases the result of such changes in geography would be to remove the higher ground to a distance from warm ocean currents, where its influence on atmospheric precipitation would be lessened. Very rich fossil floras and abundant deposits of coal are of more frequent occurrence in rocks interdigitated with marine deposits in the positive phases, in a manner suggestive of the deltaic deposits of rivers draining areas of great rainfall. Each successive phase shows marked accession in new forms belonging to more highly evoluted groups of plants, and extinction of older types. We may infer that the gradual reduction of continental areas, induced by the marine transgressions, was accompanied by increase in areas subject to a humid climate at the expense of arid regions.

3. From glaciation, large areas of the earth's surface undergoing sterilisation under ice sheets, with further changes in the vegetation of the surrounding areas from change of climate. On the melting of the ice sheet the whole area is transformed. The soils of æons have been removed, the nature of the topography has been altered, and various peculiar sterilised deposits have been plastered over the surface. The recently glaciated regions of Europe and North America exhibit large areas of peat bogs and plant formations, showing a close historical relation to the results of glaciation. More extensive areas are believed to have undergone glaciation in Permo-Carboniferous times, in an extinct continental region known as Gondwanaland, comprising parts of India and Australia, a large area of Africa, and land then existing in the present site of the Indian Ocean (8). The Permo-Carboniferous conglomerates of Australia, Africa, and India are immediately succeeded by a sequence of deposits containing a fossil flora of fern-like plants, the *Glossopteris* flora peculiar to this region and to parts of South America. It has been suggested that the appearance of this flora was due to the severity of the climate (11).

Should this view prove correct, it is probable that, as in the case of the lately glaciated districts, edaphic influences took a share in the evolution of the immigrating flora and the plant associations.

4. From volcanic action sterilising wide areas. The early plant colonisations following such sterilisations have been described by Treub in the case of Krakatoa (12), and by Grönlund for certain lava-flows in Iceland (13). The results of vulcanism in various regions during the world's history must have been stupendous. The late Tertiary volcanic eruptions of the Snake River plain in Idaho are estimated to cover an area greater than France and Great Britain combined, and still form a desert of sand and bare sheets of black basalt. These fissure eruptions have buried the topography under 2000 feet, and in some places 3700 feet, of lava, only the higher summits projecting above the volcanic floods. Fissure eruptions of the Cretaceous period in India now form the plateau of the Deccan traps over an area of at least 2,000,000 square miles to a thickness in places of 6000 feet or more (14).

5. We must also keep in mind the possible effects of the slow regional changes brought about by the agents of erosion. The degradation of mountains and the production of wide alluvial plains must, of course, be accompanied by profound alteration in the vegetation, from change of climate, from removal of barriers to plant migration, and from the transformation and translation of soils. The effects of these agencies under continued favourable climatic conditions would cause a progressive succession towards the mesophytic, as pointed out by Cowles. Where, however, from the change in geography, greater drought ensued, the vegetation might undergo retrogression, and may sometimes have culminated in desert conditions, with further destruction of the peneplain by wind erosion. Under humid climates, the continued processes of erosion and deposition on a stable crust gradually lead to the formation of deeper soils of a more and more mixed or uniform character. The rock exposures disappear, and the sorting power of the streams gradually diminishes as the base level of erosion is approached. Where, however, the climate becomes arid, erosion may

proceed further by wind, and the surface materials are again sorted into coarse and fine, sand and dust.

Secular upheaval, erosion, and deposition have doubtless been as necessary for the continued evolution of plant life on the earth's crust as the water which rains upon its surface. An area doomed to a perpetual plant-covering without erosion would in course of time be too depleted of mineral nutriment for the continued existence of the highest types of vegetation. The waste of mineral nutriment involved in failure of a complete return to the soil, and in perpetual leaching by atmospheric precipitation and drainage, must in time lead to an impoverishment which is only remediable by partial or complete removal of the soil-cap. The relations between animal and plant life would in such a case only ensure the return of plant food to the soil over ever-decreasing and limited spheres. Just as the plants' nitrogen, carbon, and water form only a part of the reserves of these substances in the hydrosphere and atmosphere, so the other mineral constituents of plant food are but a portion stolen during each recurrent epigene phase of the geological cycle of the lithosphere, represented by upheaval, decay, leaching, erosion, and sedimentation.

It is computed that nearly five billion tons of mineral matter in solution (15) are annually carried seawards by the rivers of the globe. Of this grand total one-twenty-fifth consists of potash, phosphorus, and nitric acid, the more important mineral constituents of plant food ; while if lime, the substance so essential for sustained fertility, be included, the loss amounts to nearly one-third. How much more of these substances is annually put into circulation by the plant life of the land, or is temporarily stored in the soil-cap or the ground waters, there is at present no means of estimating ; but it is evident that, in contrast to the areas subject to leaching, all large alluvial areas periodically flooded by the river waters are constantly recuperated with the food materials necessary for plant life, and that river deltas, that continue forming over a subsiding floor throughout a whole geological epoch, may be densely covered by vegetation without cessation.

Causes of extensive alterations in stable vegetation, other

than secular geological mutations, are to be looked for in the secular evolution and migrations of animals, and in the increasing specialisation and competition of the various species of plants. As a single, oft-quoted example, we may point to the profound effects on vegetation which must have followed the first introduction of social herbivorous mammals. The existing natural vegetation of New Zealand has evolved without such complications. A comparison of the vegetation of New Zealand with that of Patagonia and Australia should reveal much of interest from this standpoint.

By extensive changes such as those outlined above, parts of a tropical forest area may come to lie beyond their former districts of great rainfall, and the dominants of the forest vegetation must undergo degeneration, a desert may become partly habitable by grassland or forest, and an island flora may cease to preserve its isolation. Such marked alterations in climate or geography cause dislocations in the balance of relations between the various species of plants and animals and their surroundings, and chiefly in accordance with the rapidity of the transformation. In one direction it may lead to a keener struggle for existence, accompanied by extinction of these species less suited to the changed conditions ; in the other, with lessening competition and a wider dispersal of species under new conditions, it may lead to mutation and the establishment of new types. It is indeed conceivable that natural selection tends rather to the preservation of the type through restricting its area of migration to a certain defined habitat where only a certain defined mutation can reach maturity ; while a loosening of the reins, as it were, may allow of the colonisation of a wider range of habitats by distinct mutations with success. Further, as suggested by Kerner, the widening of the areas of distribution of related species may bring about an overlapping of such areas, leading possibly to crossing, and the formation of new types (16).

In these ways, and probably in others not indicated, not only may the various types of vegetation succeed one another in a manner entirely different from the ordinary topographic types of succession, with which to some extent we are familiar, but new types of species and plant associations may evolve in accordance with the new conditions.

PART III. THE RELATION OF SOILS TO CLIMATE AND PHYSIOGRAPHY.

Generally speaking, soil characters depend more on geography and climate than on the local nature of the underlying rocks. The soils of high altitudes, high latitudes, and deserts are usually mechanically formed and porous, and such as Thurmann designated by the name "dysgeogenous" (17), while those of low levels in humid, temperate, and tropical climates have generally suffered marked chemical decomposition, and, according to Thurmann's classification, would be eugeogenous.

Desert soils are stated by Hilgard to contain a high percentage of lime, irrespective of the nature of the district rocks (18). The abundance of lime and alkaline salts in deserts depends almost entirely on climatic and physiographic factors.

The soils of warm, humid climates undergo profound chemical decomposition and leaching. According to Russell, deep residual red soils are formed under conditions of considerable warmth and great atmospheric precipitations from rocks of various composition and origin (19).

In arctic regions the ground is permanently frozen, except during a few weeks in summer, when the surface may thaw to the depth of a few inches. Wherever humus accumulates, the soil must be for the most part acid. The production of neutral humus and nitrates, being dependent on an absence of acidity and a certain degree of warmth (20), is no doubt chiefly confined to slopes facing the sun, and to the alluvial flood plains of rivers which have their sources in more southern zones.

In the cold-temperate regions of the earth, deep residual soils are of rare occurrence owing to the slowness in the chemical operations of weathering. Large areas have been, moreover, recently glaciated, and the soils in consequence are comparatively shallow even at low levels, and often based upon peculiar sterilised deposits of glacial origin. Porous or elevated surfaces undergo leaching in cold-temperate, humid climates, since they lead to the formation of acid humus from the inactivity of the bacteria of decay, and the want of the neutralising effect of lime. These form the centres of distribution of the heath and other associations capable of growing on surfaces sterilised by acid humus (21). The further accumulation of acid humus to form peat, and the extent that the moorland and heath associations invade other surfaces less subject to leaching, depend mainly upon the climate and topography. Widespread glacial clays often have a thick covering of peat in Scotland, but support forests in other regions.

Hilgard points out that in parts of N. America the richer vegetation of the lower regions only ascends to higher altitudes in limestone districts, and suggests that the comparative fertility is associated with the presence of lime, and that the poverty of the vegetation in highland districts other than those formed of limestone is due to the leaching of the rocks and the removal of the lime in solution to lower levels (18). Thus we must consider as the effects of physiography, not only the leaching of the alkaline bases from the rocks at high levels, but also their accumulation in the ground waters at low levels. The effects of this on plant association will be guided by the physiography and the depth and porosity of the soil.

The effect of limestone on vegetation is probably much more complex in its physiographic relations than is usually assumed. Pure, soft, porous, unjointed limestones, like the chalk, undergo nearly equal surface solution over wide areas, but their success or failure in locally accumulating a leached soil depends on the purity of the rock, the climate, and the contour of the ground and the water table.

The localised solution of hard, compact, jointed limestone tends to the formation of an extensive underground drainage system wherever such a limestone country is raised above the base level of erosion. The evolution of such a drainage system depends on the persistence of a limestone area in this relation for long periods under a wet climate. It is also probable that a soil-cap with abundant humus, and somewhat permeable to water, conduces to its development. An area consisting largely of limestone rocks without such an underground drainage system may accumulate a deep soil under humid conditions favourable to luxuriant vegetation. Such a soil, especially near the surface, may contain little lime and the vegetation may form no index to the underlying limestone except by a luxuriance of deep-rooted species due to the effect of lime in the ground waters. Where, on the other hand, the climate is uniformly moist and cold, the limestone surface may become thickly covered with peat supporting a moorland flora. In either case, with the development of underground drainage channels, the ground waters will sink to a greater depth, and the soil or peat will come to suffer drought in comparison with that of the surrounding country. This will be specially marked if the area undergoes elevation relatively to sea level. The plant associations will gradually come to differ from others of the surrounding districts, not as a result of the chemical effects of lime, but from drought induced by the soluble nature of the rock in its physiographic relations to drainage. A further stage is reached when the soil or peat is gradually removed by the wind or through pipes and sinks into the underground drainage channels, and the surface is reduced to the barren condition geologically known as limestone pavement.

The actual chemical effect of lime on plant association is seen where the plants grow directly on limestone rock, where the soil overlying such rock is very shallow owing to its slow accumulation or constant removal, or where lime in solution affects the ground or surface waters.

(To be continued.)

REFERENCES.

(1) SCHIMPER, Dr. A. F. W.—Plant Geography upon a Physiological Basis. Eng. trans.
(2) GRAEBNER, P.—Die Heide Norddeutschlands. 1901.
(3) MEMBERS OF THE CENTRAL COMMITTEE FOR THE SURVEY AND STUDY OF BRITISH VEGETATION.—Types of British Vegetation. Edited by A. G. Tansley, M.A., F.L.S. 1911.
(4) COWLES, HENRY C.—"The Physiographic Ecology of Chicago and Vicinity: a Study of the Origin, Development, and Classification of Plant Societies." Botan. Gazette, 1901.
(5) CLEMENTS, F. E.—"The Development and Structure of Vegetation." Botan. Survey of Nebraska, 1901.
(6) MOSS, C. E.—"The Fundamental Units of Vegetation." The New Phytologist, vol. ix., Jan. and Feb. 1910.

(7) COWLES, HENRY C.—"The Causes of Vegetative Cycles."
     Botan. Gazette, March 1911.
(8) SUESS, EDUARD. — The Face of the Earth. Eng. trans.,
     vol. ii.
(9) PENCK, Dr. ALBRECHT.—Morphologie der Erdoberfläche.
(10) SUESS, ED.—The Face of the Earth. Eng. trans., vol. i.
     p. 387.
(11) SUESS, ED.—The Face of the Earth. Eng. trans., vol. i.
     p. 404.
(12) TREUB, M.—" Notice sur la nouvelle flore de Krakatau."
     Ann. Jard. Bot. Buitenzorg, vii., 1888.
(13) WARMING.—Œcology of Plants. Eng. trans., p. 351.
(14) GEIKIE, ARCH.—Text-Book of Geology. 1903.
(15) MURRAY, Sir JOHN.—Scot. Geograph. Mag., vol. iii., 1887.
(16) KERNER and OLIVER.—The Natural History of Plants.
(17) THURMANN, J.—Essai de physostatique appliquée à la chaine
     du Jura. 1849.
(18) HILGARD, E. W.—Soils in the Humid and Arid Regions.
     1907.
(19) RUSSELL, I. C.—"Subaerial Decay of Rocks and Origin of the
     Red Color of Certain Formations." U.S. Geol. Bull., vol.
     viii. p. 535.
(20) HALL, A. D.—For the history and literature of the study of
     nitrification, see " The Soil," 2nd ed., 1909, p. 191.
(21) GRAEBNER, P.—Die Heide Norddeutschlands.

Remarks on some Aquatic Forms and Aquatic Species of the British Flora. By Arthur Bennett, A.L.S.

DR. GLÜCK of Heidelberg is engaged on a study of the aquatic species of the European flora ; and Dr. Rothert of Krakau, Austro-Hungary, on a monograph of the genus *Sparganium.*

Both have been this year in England to study our collections, and I had the pleasure of seeing them at my house and talking these matters over with them ; these notes are a result, and we trust that our species will be examined.

To Mr. G. West's examination of about 140 Scottish lochs we are indebted for many hints as to the aquatic condition of various species, both of terrestrial, semiaquatic, and aquatic species.[1] In these two papers Mr. West gives the vegetation of these lakes (including in many instances mosses, lichens, and algæ) in their submerged, littoral, and surrounding conditions.

He discusses many subjects, and these papers, it is to be hoped, will be followed up by others, after the manner of Dr. Magnin with the French and Swiss lakes of the Jura,[2] and the United States in Bull. Michigan Fish Commission, No. 2, "The Plants of Lake St. Clair," A. J. Pieters, 1894.

1. *Ranunculus Flammula*, L., var. *natans* (Pers.).—This is a remarkable form of *Flammula* found by Mr. West in two places, a floating form at the margin of peaty pools about Morton Lochs, Tents Muir. "A strong plant 2 to 3 feet long," and a submerged form in the margins of lochs and slow streams in water 6 to 24 inches deep. Abundant in Lochs Recar, Ballochling, etc. (Kirkcudbright). Persoon describes this as "γ *natans*, fol. inferiorib. ovatis integris, superioribus linearibus," "Syn.," pl. v., ii. (1807), p. 102. Recorded from the same place by Lamarck in "Ency. Meth.," v., vi. (1804), p. 98, but given no name.[3]

2. *Ranunculus lingua*, L.—The early submerged leaves of this species, first called attention to by the late Mr. Roper,[4] are so unlike the flowering stage leaves that unless one had watched the plants it could hardly be believed ; they are 8 to 9 inches long by 3 inches wide, and in those I watched were quite decayed when the plant flowered.

3. *Peplis Portula*, L.—Mr. West found an entirely submerged form in Loch Doon (50 to 100 feet deep), Fife, "growing to a length of 3 feet with larger, thinner, semipellucid leaves, stems weak." This is quite beyond anything I have seen ; I have gathered it in Surrey (submerged) 13 inches long only.

4. *Hydrocotyle vulgaris*, L.—Usually a creeping species

---

[1] 1. "Comp. Study of Dominant Phanerogamic, etc., Flora of Aquatic Habit in three Lake Areas of Scotland," Proc. Roy. Soc. Edin., xxv., 1904-5, with fifty-five plates.
  2. "A Further Contribution," as above, Proc. Roy. Soc. Edin., xxx., 1910, with sixty-two plates.

[2] "Rech. végét. Lacs du Jura," Revue Gén. de Bot., v. 241, 303.

[3] Mr. Ewing's *natans*, "Ann. Scot. Nat. Hist.," p. 237, 1894, seems different from above.

[4] "Jour. Linn. Soc.," xxi. (1886), p. 380.

among higher vegetation in wet or damp places, but in Barlockhart Loch, Wigtownshire, Mr. West gathered a "floating form having stems 30 to 50 inches long, with leaves only ½ inch in diameter and very thin."

5. *Apium inundatum*, H. G. Reichb.—This species varies considerably as to depth of submerged forms. I have seen it in water 2 feet deep. Mr. West records it "in water from 3 feet to 6 feet deep, reaching the surface from even the greatest depth." In Engler's "Bot. Jahrbüchern,"[1] Dr. Glück throws these various semiaquatics into groups under three series (p. 140)—

1. *Die submerse form*,
2. *Die Schwimmblattflora*,
3. *Die Uferflora*,

placing this species with "*Sium latifolium, Œnanthe fistulosa, Œ. fluitans, Littorella lacustris*, etc."

Mr. West (2) remarks: "In some places, where the water has retreated, the seedlings grow so thick as to cover the mud with a sward, but their further development in an aerial environment is restrained."

6. *Ceratophyllum demersum*, L.—Not from any submerged point, but as a remarkable instance of distribution, I mention this. In the whole of the Loch Ness area, the island of Lismore, and Nairn ; then in Kirkcudbright, Wigtown, Fife, and Kinross (140 lochs), only once did Mr. West collect this species. In Otterston Loch (Fife) "it grows in such extraordinary abundance that in many places a boat can only be rowed through it with difficulty." This I have experienced in Norfolk, where in Blackfleet Broad it was impossible to force the boat through it ; it there grew intermingled with *Chara polyacantha* in thick masses.

7. *Juncus supinus*, Moench.—In this we have another species that passes from a strictly terrestrial form through many phases to an extreme one in var. *fluitans*. I have not been able to absolutely trace the submerged form to a terrestrial form, and cultivation is required here, which is just what Dr. Glück is doing at Heidelberg. We seem to possess—

1. The type as var. *nodosus*, Lange.

[1] "Uber die Lebensweise der Uferflora," 1909, pp. 104-119.

2. var. *pygmæus*, Maisson.[1]   " Caule 1–2 pollicari ; anthela depauperata 1–2 cephala."

3. var. *uliginosus*, Fries.

4. *Kochii*, Bab.

5. *subverticillatus* (Wulf.).

6. *fluitans*, Fries.

What must be a remarkable form is *J. confervaceus*, St. Leger, "Cat. bass. Rhône," 749 (1882)=*f. confervaceus*, Buchenau, in "Fl. N.W. Tiefeb." 136 (1894).

Mr. West has a very interesting note on these forms at p. 976 (No. 1). It is too long to extract, but two remarks may be quoted : " This is one of the most protean species imaginable." " These forms are of extreme interest ; in them we seem to be able to trace the phylogenesis of an extremely abundant and dominant aquatic plant ; from plastic terrestrial and subaquatic forms ; not now dominant nor abundant in this district."

From Rora Moss, Longside, V.C. 93, N. Aberdeen, Dr. Trail has sent me a specimen named var. *comosus*, Bréb. "*J. uliginosus* var. *c. comosus*. Capit. nombreux formés de feuill. sélacées en touffe seriées." [2]

Specimen from "pools near the Deveron, Banff, L. Watt sp.," is, I suppose, very near *pygmæus*, but it is 3 inches high, with setaceous stems and leaves and 2 fl. heads.

It seems now we are to accept "*J. bulbosus*, L., 'Sp. Pl.,' 1st ed., 327, 1753 ; Juncus foliis linearibus caniculatis, capsulis obtusis, 'Fl. Suec.,' 284," as the name for *supinus*.

8. *Scirpus fluitans*, L.—A remarkable form of this species was found by the Scottish Alpine Club in Lochan Bhe, near Tyndrum (822 feet alt.), 1891. A. H. Evans sp. I have seen nothing so slender as this. I do not know what is the result of growing it at the Edinburgh Botanic Garden, but I think there is no doubt it is a form of *S. fluitans*, though the name has several times been challenged. Notices will be found in the "Edin. Bot. Soc. Trans." for 1891, 1894, 1895, and 1903, p. 318.

[1] "Flora Neu-Vorpommern," p. 456, 1869.
[2] Brébisson, "Fl. Normandie" (1869), p. 336.

We now come to a series of aquatics which Dr. Glück has treated.[1]

He here treats the plants under their growth and evolution.

9. *Alisma Plantago*, L.—He divides into two varieties—*latifolium*, Kunth, and *lanceolatum*, Schultz ; again dividing these into two forms each—*aquaticum*, Glück, and *terrestre*, Glück.

The var. *graminifolius*, Wahlb., he gives as a species *A. graminifolium*, Ehrh., dividing it into four forms: *angustissimum*, Asch. et Graeb. ; *typicum*, Beck-Managetta ; *terrestre*, Glück ; and *pumilum*, Nolte.

10. *A. ranunculoides* he places under *Echinodorus*, dividing it into five forms: *typica*, Glück ; *natans*, Glück ; *zosterifolius*, Fries ; *terrestre*, Glück; and *pumilus*, Glück. He then takes the var. *repens* (Lam.) and divides that into four forms.

No doubt many of these phases of plants may be found among Scottish specimens. I have *A. graminifolium* from Perth, Dr. B. White sp. ; and Mr. West found it in Loch Gelly, Fife, as the *f. typica*.

The var. *zosterifolius*, Fries, in " Bot. Notiser " (1840), p. 35, "foliis longissimus linearibus natantibus (från Öland, Sjöshand) är en markvärdig med Al. Plantago graminif. analog form."

Mr. West gathered this in Loch Corsock in S.E. Kirkcudbright, where "it flowered under water at a depth of 3 feet ; without the flower-stalk these submerged forms look extremely like *Isoetes lacustris*." The Rev. E. S. Marshall[2] gathered it from " peat-holes above the Beauly river, E. Inverness, 1892."

Mr. West found the "*f. terrestre* at Loch Leven 1½ inches high."

11. *A. parnassifolia*, Bassi (*Caldasia parlatore*) has lately been found in Bavaria,[3] but I suppose we can hardly expect it in our isles. It has deeply cordate leaves, and only 6 to 9 carpels.

12. *Alisma natans*, L.—Are there any Scottish specimens of

[1] In " Allg. Botanische Zeitschrift " for 1906, under " Systematische Gliederung der europäischen Alismaceen."

[2] *Cf.* " Jour. of Botany " (1893), p. 48 ; and Bosch., " Prod. Fl. Bataviæ " (1850), p. 253.

[3] Glück, " Mitt. Bay. Bot. Ges." (1910), p. 285.

this species in herbaria? I have been unable to see one, but Hooker and Arnott[1] give "Black Loch, 6 miles from Stranraer," Wigtown; and in the 2nd ed. of "Topl. Botany," "Ayr, Duncan cat." is added. Mr. Scott-Elliot[2] does not notice the species at p. 164, though he gives all Wigtown species. That it is often misnamed is certain, as for some years Ireland was credited with it, but it has been found to be a submerged state of *A. ranunculoides*. It is certainly to be found in seven Welsh counties, and in two English (*i e.* Salop! and Chester!), and perhaps in two others (York and Cumberland). Dr. Glück divides this into *f. typica*, Asch. et Graeb.; *f. sparganifolius*, Fries; *f. repens*, Asch. et Graeb.; and *f. terrestre*, Glück.

13. *Sagittaria sagittifolia*, L.—Glück divides this into *f. typica*, Klinge; *f. natans*, Klinge; *f. terrestris*, Bolle; and *f. vallisnerifolia*, Coss. et Germain.

Why is *Sagittaria* so rare in Scotland? Two counties only are given, and Messrs. Kidston and Stirling added Stirling. Yet in Scandinavia it occurs from S. Sweden (Scania) up to Norland, and is not uncommon in Finland up to 67° 50′ N. lat., is absent from Finnish Lapland, but reported from Umba in Russian Lapland, and S. Norway.

Kirchner (*l.c.*) figures a curious form of *Sagittaria*, var. *Bollei*, Asch. et Graeb., which I gathered near Croydon in 1888. In this the three lobes of the leaf are only 4 mm. wide, and the basal lobes at an angle of 45°; the leaf-petioles are more succulent, and show the transverse partitions strongly when dry.

The plate (Tab. 2) that accompanies the classic account of *Sagittaria*,[3] with its details, is quite up to many of the recent drawings of aquatics, and far beyond most. The protecting sheaths of the stolons are in fig. 1 beautifully drawn.

Anyone gathering the var. *vallisnerifolia*, Coss. et Germ.,[4] might well be excused in not referring it to *Sagittaria*; the leaves are all submerged, linear, varying in length with the

[1] "Brit. Flora," 8th ed., 1860, p. 471.
[2] "Fl. Dumfriesshire," 1896.
[3] Nolte, "Bot. Bern. über Stratiotes und Sagittaria," E. F. Nolte, 1825.
[4] Cosson et Germain, "Fl. En. de Paris," ii. (1845), p. 522.

depth of water (6 dm. l), and 10 mm. wide. This I have gathered in Surrey, and seen in Norfolk.

A North-American species, *S. heterophylla*, Pursh, has established itself in the River Exe, near Exeter, Devon.[1]

Sparganium.—Dr. Rothert has found in the late Mr. Beeby's herbarium two specimens of *Sparganium* from Shetland that recede from *S. minimum* and approach *S. hyperboreum*, Laest., " Bih. i. bot. årsber," 1850. A species of N. Finmark, S. Norway, N. Sweden, Finnish and Russian Lapland, N. Finland, Iceland, Greenland, Labrador, and Hudson's Bay.

S. glomeratum, Laest., *l.c.* (*S. fluitans*, Fr.), is another Scandinavian species that should be sought for; this occurs as far south in Sweden as Scania.

In another work[2] Glück's ideas are still further worked out, and many figures (Nos. 324 to 379) are given of varying forms, leaf and other sections.

But in neither work are there any attempts to clear up or collate the many other names under these species given in European floras (mainly under *Alisma*). Neither do Ascherson and Graebner, in their review of these genera, account for many named varieties.

The following experience will show the sequence of a species of aquatic (*Damasonium Alisma*, Mill.) that does not reach Scotland, but which I watched through the summer of 1887 on a common (Mitcham) near here. It had then several ponds, many ditches, and swampy places on a gravelly soil. In one pond (since filled up) the above plant grew pretty abundantly. In April it was the form *graminifolium*, Glück; in May it began to make itself into the form *spathulatum*, Glück; at the end of June it had become the form *natans*, Glück; flowered through July and part of August; at the end of August the water became very low, and the plant here and there became stranded; it was now the form *terrestre*, Glück. The only one I could not say I saw was the form *pumila*, Glück, which he describes as " misera forma terrestris semine nata, etc."

[1] Hiern., " Exch. Club. Rep. for (1908)," p. 399, 1909.
[2] " Lebensgeschichte der Blütenpflanzen Mitteleuropas," by Drs. Kirchner, Löw, and Schröter, 1907.

The figure in "Eng. Botany," t. 1615 (3rd ed., t. 1442), shows a state between Glück's *natans* and *terrestris*. In the description no mention is made of any other leaves than the cordate floating ones. The plants noted were simply the growth and evolution of the species, influenced by warmth and depth of water. Certainly in July and August, when gathering this in other parts of Surrey, only the form *natans* could be seen on the water surface; but carefully working in the mud, the form *graminifolius* was found, probably the result of last year's seeding. Syme in "Eng. Botany" queries it as a perennial; Hooker and Babington are silent on this point. Bentham calls it an annual. Grenier and Godron[1] call it perennial, and Ascherson and Graebner[2] also. My own opinion is that it is neither, but a biennial, as I never was able to find any stolons as in *Hydrocharis* or *Alisma*; and the seeds evidently drop off, sink (they sink at once when ripe!), and in winter or early spring form the little tufts found in July with grass-like leaves.

Some Modern Aspects of Applied Botany.
By A. W. Borthwick, D.Sc.[3]

ALL progress of nations and increase in population is preceded by the discovery of some new natural resource or by a new use of a previously known one. Science is the working force which leads to increase of knowledge and industrial progress. Industries increase, but natural resources are in danger of exhaustion. The average man demands more and more, and his needs increase with civilisation and industrial progress; hence at the present day he must study, investigate, and learn how to utilise the natural resources with the greatest economy; and as man demands more from Nature, she in turn demands more from man. He must learn how to care for his crops more scientifically, to increase their yield, and also to conserve and improve the soil. In what way can the natural resources be best and most economically utilised? The obvious answer is by studying them in a scientific manner in order that we may learn how to utilise them in a scientific way. In the realms of science botany stands out pre-eminently as the science which comes into the most

[1] "Flora France," iii. (1855), p. 167.
[2] Syn. Fl. Mitteleurop.," i. (1897), p. 389.
[3] Presidential Address to the Botanical Society of Edinburgh, Nov. 1911.

intimate contact with the fundamental problems of life and living things. It is at the same time the science which lends itself most readily to practical application in many economic directions. Plants may be studied in a purely scientific spirit, that is, to increase the sum of human knowledge in the endeavour to satisfy the mind regarding the problems of life and existence. Impelled by his natural curiosity, man is always investigating, always discovering, and always discovering more to be discovered.

The study of plants and the operations of the laws of nature is of direct value as a humanising influence upon mankind; but when we can apply the knowledge so gained to some practical, economic purpose, we add not only to the usefulness but also to the dignity of the science.

Investigations pursued with a practical or economic object in view have in the past many times incidentally led to the elucidation of problems of interest in pure science, and, on the other hand, discoveries of the greatest technical importance have been made by men engaged in pure science investigation. It is impossible to separate the investigation of pure science from that of applied science, and every day the opinion is gaining ground that there is nothing derogatory to science in its application to the arts and industries. In pure and applied biological science, and, indeed, in all science, there should be a common meeting-ground between the scientist and the practical man. In forestry, agriculture, and horticulture the practical man comes daily in contact with phenomena of diverse kinds, and in time he comes to know a large number of isolated facts, the meaning of which he is apt to misunderstand or to misinterpret. Had the practical man a little more science, or if the scientific man came more into contact with him, much valuable knowledge would be gained on both sides, and much time and money saved.

This fact is clearly recognised by our Society, as the following extract from its general views and objects shows :—

" The attention of the Society is turned to the whole range of Botanical Science, together with such parts of other branches of natural history which are immediately connected with it. These objects are cultivated :

" By holding meetings for the interchange of botanical information, for the reading of original papers or translations, abstracts or reviews of botanical work, regarding any branch of botanical knowledge, practical, physiological, geographical, and palæontological, and the application of such knowledge to agriculture and the arts."

This is what led me to select as the theme of my address some practical aspects of applied botany.

In the whole range of botanical science, possibly œcology, physiology, and pathology are the three most important departments as applicable to the industries and arts.

The Society at the present time is strong in œcological experts, and has within the last year or two published several important communications bearing on œcological problems.

Regarding the introduction and cultivation of new plants of economic value, we find that this branch of economic botany has not received as much consideration and thought as its importance deserves. No doubt the world has been searched for plants of value in horticulture, and many trees, shrubs, and herbaceous plants of great ornamental value have been introduced, but no properly organised and systematic endeavour has been made to introduce and test new species of economic rather than of ornamental importance. Several of our large seed firms and many private individuals have rendered valuable service to the nation by the production of new and improved varieties of plants already in cultivation, and the recent science of Genetics is certain in the near future to enable man to produce with greater rapidity and certainty plants of improved quality. It is, however, not only necessary to discover or to artificially produce new varieties : we must go further, and test the suitability of these new species and varieties under varying œcological conditions. Owing to the endless modification in soil and climate, it is not possible to formulate definite rules, and to say whether a variety which is a success in one place will do equally well in another. Still, by making an œcological study of a plant in its native habitat, we can form a very good opinion of how it will behave when introduced into new conditions. A knowledge of a plant's œcological characteristics enables us to select those conditions of locality and environment which are most likely to supply its requirements. At the same time it is only by actual trial or experiment that such questions can be definitely settled. We must apply to nature direct for our information, and ask such questions by means of experiment, and note the reply she gives. I wish here to emphasise the fact that experiments based on scientific principles are likely to yield better and more valuable results than those conducted on blind trial and error, or rule of thumb methods. It is therefore essential that such trials should be carried out under expert supervision. It may be that some slight error in cultivation leads to failure ; hence it is necessary to know in each case when failure occurs why it occurs, and, having found the cause, to try if anything can be done to ameliorate or modify the conditions to suit the plant. Otherwise, through some initial error or failure to select the proper cultural method, a plant might be lost which would otherwise have proved a valuable addition to the economic flora. It is in connection with such problems that the study of plant œcology will prove of great economic importance. In itself the study of plant œcology, or the geographical distribution of plants on a physiological basis, is of the highest scientific value ; and when we can apply its results in a practical manner to the cultivation of plants, it assumes an economic value of equal merit.

As an illustrative example, let us for a moment consider Dr. Kienitz's important investigations into the shapes and types of the

Scots pine, as it furnishes a splendid example of the value of œcological studies to a practical industry like forestry. He has shown that the tree occurs in several œcological forms, among which two well-marked œcological types can be readily distinguished. The one a strong-branched, strong-crowned tree, which is the typical form in Scotland ; the other a slender, pyramidal-shaped tree, which is typical of the Baltic provinces. Such types are found to be hereditary, and are not altered by altered climate and soil. The Scottish type is better adapted to hold its own in the struggle for life in milder localities, whereas the slender, pyramidal type is better able to hold its own under more rigorous conditions, where wind, and especially heavy snowfalls, constitute the primary dangers. Great care should therefore be exercised in selecting seed or in planting seedlings, of whichever form is better suited to a definite locality.

In many places in this country we have examples of the Alpine or northern forms of the Scots pine planted where they have no business to be.

Plant Disease.—In applied botany there is no department of greater economic importance than that which deals with plant disease, *i.e.* plant pathology.

Plants of all kinds as living things are subject to disease. It is difficult to give a short, concise definition of disease. The following is a dictionary definition : " A derangement in the structure or the function of any organ belonging to a vegetable or animal," but this does not convey any definite or clear meaning. After all, probably it does not matter much whether we can in words draw up a hard and fast definition, as there is no hard and fast line of demarcation between health and disease. Let us for the present understand by disease any marked deviation of the vital functions from the normal. We may have various stages of healthiness, from the perfectly healthy body or organ, through less healthy conditions, till a diseased state is reached, and likewise we may have various stages of disease, from slight to severe. For health, the living body and its organs must be normal, and the environmental factors and conditions must approach as near the *optimum* in each case as possible. Otherwise, signs of unhealthiness may appear, probably not sufficient at the moment to cause much alarm to the practical cultivator, but the scientist knows that plants, even though slightly weakened, may have developed a predisposition to unhealthiness, and crops, though not actually diseased, may nevertheless be liable to an epidemic attack, the conditions for which may be made favourable through loss of vigour of the plants.

Loss of vigour and subsequent unhealthiness in plants may be caused by unfavourable environment. In the plant kingdom, as in the animal kingdom, unfavourable environmental conditions may be grouped under two heads, namely, physical and organic environment. The physical environment is supplied by the soil in all its variations of chemical composition, depth, porosity, moisture, temperature, texture, air-content, etc. Then we have the atmospheric conditions, namely, temperature, moisture, precipitations, lighting, wind, etc., the conditions of soil and atmosphere; in other words, the climate varies with latitude, altitude, aspect, and exposure. Man has but little power to ameliorate the atmospheric conditions, but he often does the opposite by allowing air to become polluted by smoke and poisonous fumes, thus producing conditions highly inimical to plant life.[1]   On the other hand, something may be done to ameliorate the soil conditions. For example, by draining, manuring, and cultivating. As regards altitude, exposure, and aspect, man can select and cultivate species or varieties in those localities or situations where each is likely to find the nearest approach to its *optimum* conditions.

The organic environment is supplied by the plant and animal kingdoms. In the vegetable kingdom we have plants in competition with one another for the best soil and air space. We have also the saprophytic and parasitic forms. The soil bacteria are not the least important members of the organic environment, although mentioned last.

The science of Mycology is of the greatest importance in pure and applied botany. Fungi play a very important rôle in nature as saprophytes and parasites. Many forms have by careful selection and cultivation been pressed into the service of man in such important industries as cheese-making, bread-baking, wine and cider preparation, brewing, distilling, etc.

The cultivated mushroom and the numerous wild edible forms, only too little known, have their importance as food plants. Finally, the study of the parasitic disease-causing forms is of the highest theoretical and economic importance.

Pure science in investigating the effect of disease aims at discovering the changes of the living substance and tissues. It may be called Cytopathology. Applied science considers the influence of disease on the plant as regards its economic value. We must combine both in order to understand the phenomenon from an economic standpoint.

The fungi, bacteria, and insect enemies of cultivated plants cause enormous damage and annual loss, not only to the cultivators of plants, but to the nation as a whole.

In Prussia the Phytopathological Commission gave in 1893 a striking example of the loss caused by grain-rust. The data were supplied by the Prussian Statistical Bureau, so that the figures are official. In 1891 the wheat harvest amounted to 10,547,168 doppelcentner, which at 22 marks per dc. would have amounted to £11,459,690, but 3,316,059 dc., or £3,595,758, fell to be deducted through depreciation by rust. From the rye harvest had to be deducted £8,896,364. Similarly, from the oat harvest had to be deducted £8,138,023. Hence the loss on a single

---

[1] The question of smoke and fume damage to plants is receiving greater attention than ever on the Continent. Such damage has increased enormously with increasing industrial development.

harvest of wheat, oats, and rye amounted to £20,628,147. In Australia the loss in wheat in 1891 caused by rust was estimated at £2,500,000.

The coffee-leaf disease of Ceylon caused by the fungus *Hemileia* was stated by Professor Marshall Ward to have cost Ceylon over a million pounds per annum for several years. He further states in his book on *Disease in Plants* that one estimate puts the loss in ten years at from £12,000,000 to £15,000,000. He further states that the Hop Aphis is estimated to have cost Kent £2,700,000 in the year 1882. If the recent outbreak of gooseberry mildew of the American type had not been scheduled under the Destructive Insect and Pest Act, and arrested, it would no doubt have wiped out the gooseberry crop throughout the country. Mr. E. S. Salmon states that the average annual value of the gooseberry crop in Kent, Wisbech, Evesham, Calstock, and Gloucestershire is from £97,000 to £160,000 in these districts alone. Also, that the value of the gooseberry crop to cottagers, private gardeners, etc., is incalculable. It was principally through his energy and influence that the disease was scheduled.

At the British Association this year Messrs. Barker and Hillier described a disease known as Cider Sickness, that causes a loss probably amounting to several thousand pounds sterling each year in the West of England alone. It is brought about by a bacterium.

A destructive bacterial disease of the banana and plantain has recently been discovered in the West Indies. The disease causes the leaves to become yellow and drop off. The terminal bud is eventually killed and the whole plant rots down to the ground. The organism responsible for this has been isolated, and is being provisionally called *Bacterium musæ*.

A disease like this might easily become epidemic and ruin the cultivation of the banana in the West Indies, in the same way as the coffee-leaf disease ruined the coffee industry in Ceylon.

I need not comment further on the loss that would result if plant pathologists were not on the spot, studying and devising means to prevent such a catastrophe.

If we had any means of estimating the loss caused annually by the dry-rot fungus, the figures would no doubt be equally astounding.

We have very little means of estimating cases of annual loss in this country due to disease, but the total must be enormous. Reference to the failure for some years and threatened extinction of the potato crop in Ireland about sixty years ago, with its attendant loss and suffering to millions of people, may recall the seriousness of an epidemic disease of a food plant. Since then preventive means in the shape of spraying have been devised, thanks to the development of applied botany, whereby the disease may be kept sufficiently in check to prevent a repetition of such a dire calamity.

Within recent years an entirely new potato disease was discovered by Schilberskzy in Upper Hungary, namely, the Black Scab disease, which is caused by a fungus, *Chrysophlyctis endobiotica*. In 1901 this disease was found in England by Professor M. C. Potter, and year by year it spread round the originally affected area, and also appeared in more distant localities, till centres of infection were reported from all over the country. In 1908 it was scheduled under the Destructive Insect and Pest Act as a notifiable disease. It is to be hoped that this may be effective in checking its further spread in this country. About two years ago this same disease was discovered by G. H. Gussow in Newfoundland. In connection with the outbreak, Mr. Gussow was sent by the Department of Agriculture of Canada to investigate the origin of the disease, and assist and advise the Newfoundland Government in dealing with it. He found the disease to be far more prevalent in Newfoundland than was supposed, and, needless to say, preventive measures have been put in operation by the Canadian Government to prevent the introduction of this disease to Canada. We can only form conjectures as to what would have been the result had this disease appeared forty or fifty years ago. Although it had not actually become epidemic, still we have every reason to suppose that in time it would have become epidemic if preventive measures had not been adopted as the result of timely warning. An epidemic of Black Scab would no doubt be much more severe than was the epidemic caused by *Phytophthora infestans*. The resting spores of the Black Scab fungus are extremely resisting to drought, and may remain capable of causing infection although kept in a very dry state for years. The disease itself is not influenced by varying seasonal conditions, such as wet and dry seasons. Hence, if it had been allowed to get the upper hand, its virulence would have been very severe. Who knows but that the disease may have been imported from this country to Newfoundland, and who knows what other British dependencies may not have been similarly laid open to the risk of infection? In any case, it is essential that all the resources of science should be employed to stamp out any new disease directly it appears.

Owing to the increased and more rapid import and export of plants and plant products (seeds, fruits, tubers, etc.) there is an increased danger of their attendant diseases being spread all over the world. In connection with disease of forest trees, I pointed out in a paper [1] dealing with the liability of the occidental and Japanese larches to be attacked by *Peziza Willkommii* that: "As regards the introduction of exotics which are intended to be grown as timber-producing trees, certain objects must be kept in view. For example, an exotic is worthy of cultivation in our forests—

"Firstly, if it is of a species at present unrepresented and capable of producing timber of utility, or if it possesses advantages as regards rate of growth, and is less exacting as regards soil and climate.

[1] "*Peziza Willkommii*, R.H., on *Larix occidentalis*, Nutt., and *Larix leptolepis*, Gord.," published in "Notes from the Royal Botanic Garden, Edinburgh, No. xxi., August 1909."

" Secondly, the introduction of an alien species is desirable if it is capable of resisting indigenous diseases, but great care must be exercised so as not to introduce a new disease along with an alien species. An exotic parasitic fungus if introduced may become rampant on indigenous species, and, *vice versa*, an indigenous parasitic fungus is equally liable to attack an exotic host. . . .

" It is therefore quite possible that exotic trees from virgin forests, when introduced into a new country and grown under artificial conditions, may readily become a prey to parasitic fungi, although hitherto in their native habitat they may have been entirely free from disease of any kind."

We are told that in its native habitat the occidental larch is not attacked by *Peziza Willkommii*, and here was an example of an exotic species becoming the prey of an indigenous fungus, or, I should rather say, of a fungus previously introduced from the Continent with the European larch. This country was in other words the common meeting-ground of an American host plant and a fungus disease from the Continent.

As I had already learned from my former teachers in Munich, Professors R. Hartig and von Tubeuf, that such dangers existed, I was glad to be able to add this example as a warning in this country. How little such warnings are sometimes heeded here and elsewhere the following note by Professor von Tubeuf in his Journal, entitled " Naturwissenschaftliche Zeitschrift fur Forst- und Landwirtschaft," will show. He says : " In an article published in the Year-Book of the German Dentrological Society for the year 1904, p. 156, I drew attention to the danger and frequency of the spread of plant parasites by commerce, not only within a country, but from one country to another, and even to distant parts of the world. I also cited several instances of such occurrences. The transmission of the rust disease of the Weymouth pine within Germany by young infected plants was a typical example.

" The news which now comes from America is still more interesting and significant. This dangerous disease of the Weymouth pine has been introduced into America, the home of *Pinus strobus*, with a consignment of seedlings sent from Hamburg. In America the disease was unknown and had never been seen there on native trees. Every precaution has been taken to prevent the spread of the disease, and it is hoped that this invasion may be repulsed as a previous appearance of the disease in New York in 1906 was immediately stamped out.

" It is difficult to understand why America imports seedlings of *Pinus strobus* in spite of all European experience and warning, instead of supplying her own wants by seedlings raised from native seeds."

Here is a remarkable example :—An American tree is introduced into Europe, becomes the victim of a European fungus, and in course of time young diseased plants are sent from Europe to America, to the imminent danger of the indigenous trees. The loss to America would have been great if she had not had an organised department to arrest and quarantine these infected plants even at the eleventh hour.

The Weymouth pine is a very desirable tree to grow, not only arboriculturally for ornamental purposes, but also sylviculturally for the sake of its timber. It is called White pine in America, but on this side of the Atlantic its timber is known as Yellow pine, a kind of timber which has become very scarce and expensive of late years, owing to its having been too severely exploited in America. I have seen this tree growing well in the South of England, where it gave every promise of forming an excellent stand of timber in a comparatively short rotation, but I also noticed traces of this disease in its neighbourhood. In Scotland the Weymouth pine grows quite well in suitable places, but the disease is unfortunately rampantly epidemic.

The fungus belongs to the group of metoxenous forms, its alternative host being almost every species of Ribes—certainly *R. nigrum*, *R. alpinum*, *R. aureum*, and *R. grossularia*. This fungus is doubly injurious, since it attacks two host plants of economic importance. No effort should therefore be spared to prevent the further spread of this disease or to stamp it out, and such is not beyond the power of properly organised practical mycology. The remedy recommended is to remove whichever of the host plants is considered to be of the least economic value.

There exist among cultivated plants different varieties, some of them predisposed to disease, others immune. The immunity may be due to anatomical or physiological differences. Whatever the cause of the immunity may be, we can always test whether it exists or not by experimental methods. The fact of great importance is that it is possible to produce varieties which can resist certain diseases, and we are now learning more about the laws which govern the production of varieties, so that the special variety desired can be produced with greater certainty and rapidity than was formerly the case. Our future efforts in stamping out disease *must* be concentrated more on the rearing of resistant varieties than has been the case in the past, and this is another of the ways in which the modern science of Genetics will prove of great value in applied botany.

The remedies for plant disease are mostly all of the nature of antiseptics or fungicides. They are not of the nature of medicines, as generally understood in animal ailments. Still, much may be done by keeping the plant healthy, and supplying it with the right kind and amount of food. Attention to the proper supply of water, heat, and light is also of importance. In other words, keep the plant in a proper hygienic condition, and it will, like the animal under similar conditions, be better able to resist all kinds of disease. In order to do this we must study and understand the inter-relationship between the plant and its surroundings, and it is in this connection that the study of plant œcology from a physiological point of view is of such vital importance.

The method of dealing with outbreaks of disease when they occur locally may be of advantage locally, but individual or isolated action, though of use, is of very little avail in stamping out an existing or preventing a threatened epidemic, because the methods employed are not fundamental, they do not strike at the root of the disease.

In stamping out disease we must have properly organised and combined action, otherwise the best efforts are bound in the long-run to prove futile.

The Americans were among the first to realise the importance of plant disease as a national economic question, on account of the enormous loss which, we have seen, may be caused thereby, and they have a special phytopathological section in their Department of Agriculture, which Department dates from 1889 and has a Cabinet Minister at its head. The aggregate appropriation since 1900 is £18,002,412, and this year the Department has at its disposal about £4,000,000, double of what it was in last decade. The Department employs a staff of 12,480 men and women, 600 to 700 of whom are engaged in scientific research. The money appropriated for the Department in all its branches of activity would amount to £4,514,003. In spite of the magnitude of this sum, it is regarded in America as an investment, and not an expenditure.

An interesting item is the vote of £1000 for the study of the Chestnut Bark disease.

The Chestnut Bark disease is caused by a fungus, *Diaporthe parasitica*, a wound parasite which attacks the main trunk or branches of old and young trees. It first attacked the native American chestnut, but it has spread to the Japanese chestnut and other varieties. This disease was first discovered by Dr. Murrill in New York Botanical Gardens in 1905, and reported on by him in 1906. It spreads with great rapidity.

The chestnut trees of Greater New York have all been attacked and practically destroyed. Many valuable trees have been destroyed in all the counties of New Jersey. It has gone through Connecticut westward to the Berkshire Hills, and has spread over Long Island and Staten Island, and has reached far enough west to invade a large area in Pennsylvania. Unless some means is found to arrest the disease, it bids fair to ruin the growth of chestnuts in America, where the timber is highly prized for railway sleepers and posts, mining timber, and farm purposes. In rough construction it is used extensively. Government reports show that the yield in 1907 was 650 million feet B.M., of an estimated value of $11,000,000. The quantity used for railway ties alone amounts to $3,000,000 per year. The "Gardener's Chronicle of America" concludes an article on this disease as follows :—" The loss upon which it is most impossible to estimate in dollars is the loss to tree lovers and tree owners, who would not take any amount of money for the stately forest veterans which have been the pride of their estates. There are large areas of territory not yet reached by the fungus. We may hope that its course may be run, or that study and experiment may evolve an effective remedy before the disease shall cover the remaining chestnut territory."

The expense of the Department may be great; but when the value of the work done and research carried out in the realms of applied botany is set off against it, the expense becomes dwarfed to vanishing point. Surely the American Government is very wisely and well advised in spending a paltry thousand pounds in order that, by study and experiment, an effective remedy may be evolved to save not only the remaining chestnuts but the future crops, whose annual value is reckoned in millions.

Germany has now also a vast and well-equipped organisation for research in connection with plant protection.

At home the practical cultivator of plants may get scientific advice from the Board of Agriculture, apart from which he has to seek his scientific advice from the staffs of our Universities and Agricultural Colleges. Many private societies and even individuals have hitherto done much to disseminate scientific knowledge of great importance to the practical grower of plants. We need only glance at the publications of many different societies to find much information, not only of great scientific interest, but with a direct bearing on questions of practical importance and utility. The Agricultural and Horticultural Press has become a valuable national asset owing to the way in which it has kept abreast of the times, and disseminates a great amount of accurate and valuable information and advice on matters where applied botany can be of assistance to farmers, gardeners, fruit-growers, and all who cultivate the soil.

How the Botanical Society might extend its Range of Usefulness.

In organising their Department of Vegetable Pathology, the Americans found it necessary to have special agents in conneetion with each institution. These agents are appointed in as many different localities as possible. One of their principal duties is to instal and supervise experiments of various kinds. Their reports are sent in to the Central Institution, and when the results from a sufficient number of localities seem to justify any conclusion being drawn as to the success of, say, some preventive or remedial measure in connection with a certain disease, then such information is printed in bulletin form and circulated.

In Germany we find much the same kind of organisation. In Bavaria, for example, the Central Institution, namely. the Institute of Agricultural Botany, is situated in Munich, and is under the direction of Dr. Hiltner, who has a staff of highly trained experts in all branches of plant protection. In conjunction with this Central Institution there are a great number of local stations (Auskunft-stellen) situated in the smaller towns, villages, etc., throughout Bavaria. In charge of these centres are local agents or representatives, not necessarily expert plant pathologists, but nevertheless men of scientific training.   Many of these local representatives are clericals, schoolmasters, leading agriculturists, etc.   Their principal duty is to send in reports on specially prepared schedules to the Central Institution as nearly as possible every four weeks.   Each local station has in turn correspondents (Vertrauensmänner) in as many country districts as possible.   These correspondents are, so to speak, the men on the spot, and they are constantly in touch with the local agents, and thus the Central Institution is kept constantly informed regarding the state of field and garden crops all over the country, and, should occasion arise, experts can at once be sent to investigate and advise.   The Imperial Biological Institution for Agriculture and Forestry situated at Dahlem, near Berlin, is the principal institution, and is kept posted up to date from all the other Central Institutions of the empire.

The ceaseless activity of these institutions has already resulted in the accomplishment of an extraordinary amount of useful work of the highest scientific importance and economic value.

As a Society, we cannot hope to deal with problems in applied botany on the same scale as a well-organised and subsidised State Department, but, nevertheless, we could be of some use I think. We have Local Secretaries and members almost all over the country, and no doubt many would be quite willing to make systematic observations on the occurrence, spread, and severity of plant diseases in the forest, field, and garden of their own particular area. Doubtful cases of any disease, the cause of which was not evident to the local representative, might be sent in to headquarters, where a Special Committee might investigate and report.   The reports and records of the Local Agents and those of the Central Committee would in time become not only of scientific interest but of great practical utility.

The making and recording of these observations may seem all very simple and such as anyone might be able to make, but I do not propose for one moment that such records should contain a mere list of parasitic fungi found from year to year; such lists alone would have very little scientific or practical value.

These investigations could only be carried out by botanists. They would have to be of the nature of an œcological study of the disease.   Such factors as the influence of the soil, the climatic influence, the local method of cultivation, the nature of the attack— slight or severe—the presence of other plants, in fact all the conditions in the physical and organic environment which influence the relationship of host and parasite would be noted and recorded.   It is only by such means that we can gain any clear and definite knowledge of the conditions in nature which influence the increase or decrease of disease.   It is only when we are in possession of such records that prophylactic measures can be evolved, and plant hygiene placed upon a sound scientific basis.

The complete study of a plant disease may be presented as follows:—First, we should learn to diagnose the disease from its outward visible effect on the plant. This may be called the symptomatology of the disease. Then comes the study of the ætiology, or the investigation of the cause. Then, after the cause is known, we are in a position to find out the cure and future prevention, namely, the therapeutics and prophylaxis. The first and last of these, namely, the diagnosis and prophylaxis, are the most important from the economic standpoint.

Everyone interested in the cultivation of plants should endeavour to make himself familiar with the appearance and effect, and possibly the botanical names, of the commoner disease-causing fungi, also the general preventive measures to be adopted to prevent their spread. It is of the highest importance that a plant disease of any kind should be recognised in its earliest stages, as it is then that its spread may be prevented either by a timely spraying or by the more drastic method of removing the diseased individuals and burning them to make sure the disease-causing fungus is destroyed. It usually happens that before the advice of the plant pathologist is sought, the disease has made itself strikingly apparent by the amount of damage done. It is then often too late to effect a cure.

On broader lines these remarks which apply to the individual apply equally to the State. We have seen how the Governments of other countries have established Departments to watch over the health of cultivated plants, and they are ready at a moment's notice, so to speak, to aid these Departments by special legislation should occasion arise in the shape of a threatened epidemic.

True, our own Government has passed special Acts with the view of preventing epidemics, but, unfortunately, these special laws have been so tardy and so long delayed that their effect on the disease may be the same as the proverbial delay in locking the stable door, and in any case there is not sufficient supervision to ensure that these special laws are carried out so as to be of real value.

From the earliest times we have records that cultivated plants were subject to blights, pestilence, and disease, which the earlier cultivators of the soil attributed to various causes (moon, stars, etc.), but we also find the weather, climate, and soil held responsible for various brands, rusts, and cankers. The existence of parasites or the phenomenon of parasitism among plants was undreamt of. Still, we have here a foreshadowing of the study of the effect of the physical environment on the health of plants. We know now that certain kinds of weather and climatic conditions predispose plants to certain kinds of disease, whose life histories we know, and we also know that their relationship to their host plants is regulated by external physical conditions which may render the host plants more vulnerable, and thus enable the parasite to attack and cause disease. The weather as a physical factor may predispose plants to certain organic diseases, so that the observers in these early times were quite correct in their observations, but their conclusions were inaccurate or incomplete.

Bacteriology has mainly owed its development to the work and research carried out in connection with pathogenic forms. The importance of this department of applied botany is too obvious to require more than a passing comment. Although the layman may be accustomed to think of all bacteria as harmful, still the great dependence of the higher forms of plant life on those lower organisms is being made clearer every day. As I have already said, as man becomes more exacting upon the natural resources, so must he in turn endeavour to help Nature by artificial means. As cultivation becomes more intensive, the more must man employ scientific methods to conserve and improve the fertility of the soil, and, in this connection, valuable service has been rendered to agriculture, forestry, and horticulture by the botanist and chemist. The study of plant chemistry and plant physiology has opened up a wide field of research, in which already great progress has been made, many valuable results achieved, and probably nowhere with greater success than in the study of the soil bacteria. The rôle played by the nitrogen-fixing soil bacteria is becoming better understood every day. Successful experiments have been carried out in artificially inoculating the soil with these important organisms. In other words, the soil may be sown with these useful organisms, and upon the success of the development of this invisible soil flora depends the success of crops of higher plants. As we find almost always in Nature, these useful soil organisms have their enemies in other soil micro-organisms, and it has been found that by partially sterilising the soil, crops are improved, the improvement being due to the removal of those organisms inimical to the useful ones. In other words, methods of plant protection may be applied to protect these invisible plants, and thereby improve the quality of the soil and sustain its fertility.

During the last twelve years it is estimated that the agricultural produce of America has amounted to £16,000,000,000. The area of land under cultivation has not increased anything like so rapidly as the value of the agricultural produce ; for example, ten years ago the farm products were valued at £800,000,000, now they amount to £1,800,000,000. The increase is attributed entirely to the better and more up-to-date scientific methods generally employed by the farmers, and the change has been brought about by the Agricultural Department.

Recently, on the recommendation of the Development Commissioners, the Treasury has sanctioned the allocation of funds to be administered by the Board of Agriculture in initiating and organising schemes for systematic research in agriculture. The sum to be expended when these schemes are in full working order will be about £50,000 per annum.

Grants will be made for research in various groups of subjects, among which we note plant physiology, plant pathology, and mycology, plant breeding, and fruit growing, including the practical treatment of plant diseases, plant nutrition, and soil problems. A fund not exceeding £3000 per annum will be available for assistance in special investigations, for which provision is not otherwise made.

The Board thoroughly realises the importance of having none but carefully trained men for work in connection with the scheme. The Board therefore proposes to offer for 1911, 1912, and 1913 scholarships of the value of £150 tenable for three years. These scholarships will be twelve in number, and will be awarded only to thoroughly suitable candidates. Grants will also be made to Teaching Institutions, Universities, Agricultural Colleges, etc., in England and Wales. These will act as centres where farmers may apply for scientific advice on important technical questions, and further special investigations of local interest can be carried out by these institutions.

By means of those grants for research scholarships, local advice, and investigations, it is hoped to provide an expert staff with both scientific and practical qualifications, the members of which will be engaged in solving problems of local importance and endeavouring in every way to secure the application of science to practice.

It is very satisfactory to know that steps are being taken to provide more practical training for the university and college student in order to promote the application of science to practice. But, as I have already indicated, the practical man should have better facilities for acquiring a knowledge of the fundamental scientific principles upon which his practice is based. With one or two notable exceptions, it is not possible for the young gardener or forester in training to attend systematic courses of instruction in the sciences underlying his future profession, unless he happens to be fortunate enough to be employed in some nursery or private garden in or near large towns. This difficulty might be got over by providing bursaries to enable such men to attend courses of instruction at suitable institutions; but to this method there is the very serious objection that very few colleges are able to provide practical training under proper supervision and control along with scientific instruction; however, there are indications that this unsatisfactory state of affairs will soon be improved.

The point is—it is of importance that these practical men should not, when getting theoretical training, lose touch with practice; also, that for the practical work which they perform during theoretical training they should receive payment adequate as a subsistence allowance. A scheme whereby this is achieved is that which has been in operation in the Royal Botanic Garden, Edinburgh, during the past twenty years. Young gardeners and foresters are taken on the staff and receive certain payment for their services, and at the same time are taught free of charge the scientific element of their work. There is no place in this country where more has been done to provide employment, combined with practical and scientific training, for the young gardener and forester, than at the Royal Botanic Garden, and we should feel proud that Edinburgh has led the way in this important development; and experience has shown that the men so trained have had no difficulty in obtaining the best appointments both at home and abroad. With such provision for the training of the scientist and the practical man in applied science we may look forward to a time when the natural resources will be more carefully conserved and utilised, and the only way whereby this end may be achieved is through sound science to good practice.

Alien Plants.   By James Fraser.

THE following list of plants which I met with during 1911, unless where otherwise indicated, consists of ten new British records (these are marked with a star), several new county records, and plants mainly from localities which indicate a wider and perhaps widening range of distribution for them in counties where they have already been known to exist.

My best thanks are due to Professor Hackel, and to the Director of the Royal Gardens, Kew, for their help in determining several of the more difficult species.

Actæa spicata, L.—In the grounds of Monteviot, Roxburghshire ; several.

Matthiola tristis, Br. — Near Musselburgh, Midlothian ; several.

Silene inaperta, L. — Near Musselburgh, Midlothian ; several.

Lychnis Preslii, Sekera.—Near Tantallon, East Lothian ; one clump.

Oxalis corniculata, L. —On the shore of Loch Ryan, nearly a mile north of Sheuchan Mills, Wigtownshire ; two or three.

Staphylea pinnata, L.—In the grounds of Prestonhall, Midlothian ; several seedlings.

Trifolium Michelianum, Savi.—On reclaimed ground at the Esk mouth, Midlothian ; one plant.

Rubus spectabilis, Pursh.—Along the Heriot Water, at Borthwick Hall, Midlothian ; very plentiful.

R. odoratus, L.—In the grounds of Prestonhall, Midlothian ; several.

Poterium canadense, A. Gray.—On the shore north of Port-patrick, Wigtownshire ; several.

Sedum Lydium, Boiss.—By the roadside near Clovenfords, Selkirkshire ; plentiful.

*Crucianella patula, L.—Near Musselburgh, Midlothian ; several.

*Millotia depauperata, Stapf, sp. nov.—A plant found on the Tweed shingle at Galafoot in 1908 has been so named by Dr. O. Stapf, Keeper of the Herbarium at the Royal Gardens, Kew, the full description appearing in the *Kew Bulletin*, No. 1, 1910.

*Centaurea Moschata, L.—In an old quarry near Slate-ford, Midlothian ; several.

Crupina vulgaris, Cass.—Near Musselburgh, Midlothian ; several.

Campanula persicifolia, L.—Near Musselburgh, Midlothian ; a large clump.

Nymphoides peltatum, Rendl. and Brit.—In the pond, Prestonhall, Midlothian ; plentiful.

Verbascum virgatum, Stokes.—At Murieston, West Lothian, in 1910, and at Leith Docks in 1911 ; three or four in each.

*Phalaris truncata, Guss.—Leith Docks ; one plant in 1910. Named by Professor Hackel.

Piptatherum multiflorum, Beauv.—Near Musselburgh, Mid-lothian ; several.

Calamagrostis epigeios, Roth.—Near Portpatrick, Wigtown-shire ; one clump.

Aira provincialis, Jord.—At Leith Docks in 1905 and 1911 ; several.

Wangenheimia Lima, Trin.—Near Musselburgh, Midlothian ; plentiful.

*Avellinia Michelii, Parl.—Near Musselburgh, Midlothian ; two plants. Named at Kew.

Desmazeria sicula, Dum.—Near Slateford, Midlothian, in 1907, and at Murieston, West Lothian, in 1910 ; two or three in each.

Poa Chaixii, Vill.—Near Monteviot, Roxburghshire, in immense quantities ; in the grounds of Darnhall, Peeblesshire, in large quantities ; in a strip of wood east of North Berwick, East Lothian, scarce.

Cutandia incrassata, Salz.; *C. divaricata*, Desf.; *Festuca clavata*, Moench (*Vulpia geniculata*, Link.), all identified at Kew; and

F. tenuiflora, Schrader (*Nardurus maritimus*, Murb.), named by Professor Hackel; near Musselburgh, Midlothian; several of each.

Bromus rubens, L.—On reclaimed ground at the Esk mouth, Midlothian; several.

B. squarrosus, L., var. *villosus*, Koch; and *Brachypodium ramosum*, R. and S.—Near Musselburgh, Midlothian; several of each.

Carex helvola, Blytt. By Arthur Bennett, A.L.S.

CAREX HELVOLA, Blytt, ap. Fries, " Bot. Notiser," (1849), p. 58.

"Spica composita, spiculus subquinis linearibus confertus (distinctus imbricatis), conformibus, distigimatibus, terminali basi mascula; bracteis evaginalus, membranaceis, brevissimus; squamis ovatis, acutis; fructibus basi cuneata ovatis, glabris, in rostrum compressum subinteginus attenuatis."
—"Bot. Not.," *l.c.*; Fries, "Herb. Nor.," fasc. xiii., No. 85, 1849.

C. canescens × *lagopina*, Kihlman.

C. curta × *lagopina*.

C. curta × *Lachenalii*, "Flora Danica," Supp., t. 32, 1853.

In the Transactions of the Society in 1886 I first recorded this as a Scottish plant, gathered by Professor J. H. Balfour in 1846 on Lochnagar.

In 1897 Mr. Druce gathered it on Ben Lawers in Perthshire (14), and published an account of it (15).

In 1906 the Rev. E. S. Marshall (18) gathered it on Lochnagar. " In the great corrie which faces north (above Loch-an-Ean) it grew in wet ground associated with *C. canescens* var. *fallax*. A fine patch of *C. Lachenalii* (*lagopina*) grew within 20 to 30 yards of it, so the hybridity was easy to account for. Alt. 3500 feet (27/7/1906), in an alpine bog associated with *C. rariflora*, Sm., above the same corrie, in much wetter ground, so that the specimens were on an average considerably more luxuriant. I believe the second station to be identical with Crawford's ; E. F. Linton grew his plant, and it remained quite sterile and unchanged. There I could see neither of the parents, but I believe the original station for *lagopina* was very close at hand" (Marshall *in litt.*).

To Mr. Marshall's (R. No. 2980) specimens Herr Küken-thal has added "very characteristic." Good cultivated specimens were issued by Rev. E. F. Linton in 1909, through the Watson Botanical Exchange Club. In 1898 and 1899 Mr. Druce (19) gathered specimens on Ben Lawers that differed, and caused him to write "*helvola* var." (19). Some of these Herr Kükenthal cites under " *C. canescens × stellu-lata = C. biharica*, Simonk." ;[1] others as " *C. tetrastachya*, Traun.,[2] *supercanescens*, Kükl.."

I have specimens gathered on Ben Lawers (19/8/1886) by Messrs. H. and J. Groves which seem to be referable to *helvola*, at that date of course under " *C. curta* v. *alpicola*."

I have seen specimens also from Forfar (Edinb. Herb.), and Mr. N. E. Brown wrote (30/3/1886) that he believed there is a specimen from Ben Muic Dhui in the Kew Herbarium, but had not dissected it; and the late Mr. Beeby wrote: " I seem to have *Carex helvola* from Glas Mheil. leg. Duthie, 1874." This is in Forfar, but on the confines of Forfar, Aberdeen, and Perth, alt. 3502 feet. *C. Lachenalii* (*lagopina*) occurs in Forfar (Ewing). The fullest account we have of the plant is furnished by Herr Kihlman (6), where he also describes a hybrid between *canescens* and *norvegica*, Willd., as *C. pseudo-helvola*.

In Europe *C. helvola* occurs in Finnish and Russian Lap-land north to 69° 50′ N. lat.; Swedish Lapland; Sweden in Bohuslän, Upland, Södermanland, and Vesterbotten; Nor-way to 71° 3′ (Norman),[3] and distributed from the Birch to the Willow Belts (4); Iceland. Asch. and Graeb., " S. Mitt. E. Fl.," 1902, p. 64. Greenland (5 and 9).

I here confine myself to the original *helvola*, the Tirol, etc., plants being *C. tetrastachya*, Traun.

---

[1] "Enum. Fl. Trans.," 1886, p. 548.
[2] Sauter, " Flora," 1850, p. 366.
[3] " Not. summ. concep. Arctic Norway," 1881, p. 500.

Neuman (11) strangely puts *helvola* as "*canescens* × *norvegica*," yet has a *canescens* × *lagopina*, and refers to Hartman[1] (but he does not give *helvola* as a hybrid, merely saying "habit of the foregoing," *i.e. C. micustachya*, "Ehrh. Hann. Mag.," 1784, p. 9). Dr. Williams (17) gives other hybrids with *canescens* in Europe ; of these *C. paniculata* ×, *C. paradoxa* ×, *C. remota* ×, and *C. dioica* × are the only ones that can occur in Britain.

Mr. Fernald[2] says his *C. elachycarpa*, figs. 133, 134, "at maturity strongly suggests the little known *C. helvola*, Blytt, which, however, has very different perigynia." But the figure rather suggests a small or starved *C. curta* than *helvola*.

C. helvola seems to have been first gathered in 1826 by Holmgren (2), and by Blytt in 1833 (2), but not published until 1849.

One difficulty is, *C. lagopina* is not known on Ben Lawers.

REFERENCES.

(1) ANDERSSON.—Cyper. Scand., 1849, p. 61.
(2) BLYTT.—Norges Flora, 1861, p. 188.
(3) NORMAN.—Soc. Nat. Spec., 1864, p. 43.
(4) BLYTT.—Norges Fl., Supp., 1877, p. 1254.
(5) LANGE.—Medd. Groenland (3), 1887, p. 288.
(6) KIHLMAN.—Med. Soc. Fl. et Fa. Fenn., 1889, p. 16.
(7) Herb. Mus. Fenn., 2nd ed., 1889, p. 16, 125.
(8) BLYTT.—Ny. bid. i. Norge, 1892, p. 15.
(9) ROSENVINGE.—Medd. Groenland, Supp., 1892, p. 719.
(10) HJELT.—Fl. Fennica, 1892, p. 256.
(11) NEUMAN.—Sveriges Flora, 1901, p. 711.
(12) A. BENNETT.—Trans. Bot. Soc. Edinb., 1886, p. 361.
(13)   ,,   Jour. Botany, 1886, p. 149.
(14) DRUCE.—Ann. Scot. N. Hist., 1897, p. 260.
(15)   ,,   Jour. Linn. Soc., 1898, p. 157.
(16)   ,,   Jour. Botany, 1898, p. 157.
(17) WILLIAMS.—Jour. Botany, 1908, p. 369.
(18) MARSHALL.—Jour. Botany, 1908, p. 108.
(19) DRUCE.—Ann. Scot. N. Hist., 1909, p. 238.

[1] "Skand. Fl.," 11th ed., 1879, p. 471.
[2] "Proc. Am. Acad. Arts and Sc.," 1902, p. 492.

Ecological Terminology as applied to Marine Algæ.
By N. Miller Johnson, B.Sc., F.L.S.

THE works of Warming and Borgesen are eloquently suggestive of what research methods *should* be in ecological botany.

That of the former is magnificently comprehensive, covering as it does the whole field, while the latter is no less comprehensive, from, however, a more restricted standpoint.

Algæ are *included* in Warming's work, while they occupy the entire theme of Borgesen's " Faeroese Algæ."

No doubt the general concepts of formation and association possess similar values in the mind of each writer, but as regards Algæ the terminology clearly shows that the concepts are different.

It would appear that formation as used by Borgesen, implies association as used by Warming.

According to the former, the word formation is used to denote a group of different species belonging, as in the Fucaceæ formation of a sheltered coast, to the same family; whereas, if one correctly interprets the latter, the word formation is used to designate the entire group of Algæ (limno- or halo-nereid formation according to the fresh- or salt-water habitat).

Borgesen's association appears to be a formational unit consisting of one species only, while Warming's association seems to imply a group of plants of one, two, or more species all growing together under the same or similar conditions; or, in his own words, which must be taken to refer to terrestrial vegetation only, "an association is a community of definite floristic composition within a formation" (p. 145).

Thus it will be seen that while under certain circumstances the idea' of association as used by both writers is the same, yet in the majority of cases formation, as used by Borgesen, means association as used by Warming, and the association of the former is the plant society of Moss (p. 48, 1910).

The suggestion which the present paper wishes to embody is, that as ecological terminology is now fairly definite, and accepted as such at least in Great Britain, an effort should be made to use the same terms, if not to the entire range of cryptogamic botany, at any rate to marine algæ.

Just as a terrestrial formation may be divided into two or more sub-formations, the nereid formation (of Algæ) is divided by Warming into two sub-formations :

(*a*) Fresh-water (limno-nereid).

(*b*) Marine (halo-nereid) (p. 169).

It is customary to distinguish in the latter sub-formation two regions :

(*a*) the littoral ; (*b*) the sub-littoral.

These regions could then be again divided into associations and plant societies, according to groups, single species, or in many cases successive storeys, but it will be seen that the sub-littoral habitat would be a great drawback in actual delineation of group boundaries.

Borgesen's methods of nomenclature will be seen from the following (p. 711):—

Of the formations of exposed coasts (littoral region), he recognises among others a Hildenbrantia formation consisting of *Hildenbrantia rosea, Ralfsia verrucosa*, and blue-green Algæ; a chlorophyceæ formation consisting of *Prasiola crispa, Enteromorpha intestinalis*, and *Rhizoclonium riparium*; and a porphyra association found under the chlorophyceæ formation, and consisting of *Porphyra umbilicalis*

It will be recognised that, according to the preceding matter and suggestions, the terminology would be as under:—

The littoral region of exposed coasts being part of the general halo-nereid sub-formation, consists of, among others, a Hildenbrantia association composed of small or large plant societies of *Hildenbrantia rosea, Ralfsia verrucosa*, and blue-green Algæ; a chlorophyceæ association consisting of plant societies of *Prasiola crispa, Enteromorpha intestinalis*, and *Rhizoclonium riparium*; and, existing beneath the chlorophyceæ association as a lower storey, a plant society of *Porphyra umbilicalis*.

The following summarised notes may illustrate further the suggested method of treatment.

These brief notes are from observations made by the writer on the marine Algæ of the Kirkcaldy district of Fife (littoral region).

Nereid formation of Algæ; marine (halo-nereid) sub-formation; littoral region.

District I.—Ravenscraig to Craig Endle; rocks and sandy beach; creeks and isolated rocks.

ALGÆ.—(1) Fucaceæ association, with *Fucus vesiculosus* (minus bladders); dominant between H. and L.W.M.

(2) Plant societies of—

    (a) *Callithamnion scopulorum.*—Extensive, existing beneath *F. vesiculosus* as a lower storey.

    (b) *Enteromorpha compressa.*—Not plentiful, on rocks both at H. and L.W.M.

    (c) *Porphyra laciniata.*—Fairly plentiful at L.W.M.

District II.—Craig Endle to Dysart harbour; A, a series of creeks; B, a bay with rock areas.

A. (1) Fucaceæ association; *F. vesiculosus* (with bladders where on creek sides) dominant.

(2) Plant societies of—

    (a) *F. serratus.*—Common.

    (b) *F. spiralis.*—Near H.W.M.; scarce.

    (c) *Pelvetia canaliculata.*—In honeycombed rocks near H.W.M.; scarce.

    (d) *Enteromorpha compressa.*—Abundant; due to sewage.

    (*e*) *Laminaria digitata*, *L. Saccharina*, and *Chondrus crispus* in tidal pools.

    (*f*) *Ceramium rubrum* (pale brown), and *Enteromorpha compressa* (colourless), in, and due to high stagnant pools.

  B. (1) Fucaceæ association; *F. vesiculosus* and *F. serratus* dominant.

  (2) Plant societies of—

    (*a*) *Ascophyllum nodosum.*

    (*b*) *Callithamnion scopulorum* and *Gigartina mamillosa.*—Abundant as a lower storey on sea-wall between H.W.M. and half-tide mark.

    (*c*) *Chylocladia articulata* and *Gigartina mamillosa.*—Abundant as lower storey on sea-wall between half-tide mark and L.W.M.

    (*d*) *F. spiralis.*—Near H.W.M.; scarce.

Indubitably supporters of the view that this terminology, as applied to marine Algæ, is capable of improvement, will not be lacking.

The writer, however, ventures (with all due modesty) to suggest that the value of the suggestion lies in the fact that, if acted upon, the study of Algæ from the present standpoint will be brought into unity and conformity with existing methods of phytogeographical terminology.

LITERATURE.

(1) WARMING, E., assisted by VAHL, M.—Œcology of Plants: an Introduction to the Study of Plant-Communities, Oxford, 1909.

(2) BORGESEN, F.—The Algæ Vegetation of the Faeroese Coasts, with Remarks upon the Phytogeography. 1905.

(3) MOSS, C. E.—The Fundamental Units of Vegetation; reprinted from New Phytologist. 1910.

Short Notes.

[It is hoped that all will combine to make this section as complete as possible by the prompt recording of all " new records," etc.]

Barbula gracilis Schwaeg. New to Scotland.—In October last year I gathered in Glen Phee, Clova, *Barbula gracilis* Schwaeg. This moss, although unrecorded for Scotland, is probably often overlooked owing to its resemblance to *Barbula fallax*. It is rather interesting to note that it grows on the same rock faces of quartzose gneiss as *Oxytropis campestris* and *Sælania cæsia*. The moss, as is always the case in Britain, was sterile. Specimens were sent to Mr. D. A. Jones, Harlech.   MARGARET CORSTORPHINE.

Centunculus minimus, L., in Wigtownshire.—In August last I found this plant growing in small quantity on Craigoch Moor on the footpath on the top of the cliffs approaching Morroch Bay from Portpatrick; a new record for vice-county 74.   JAMES FRASER.

The following are a few new records for Scottish vice-counties :—

Silene fimbriata, Sims, from the south bank of the Crinan Canal, Argyllshire, at Auchendarroch Lodge : evidently an outcast.

Piptatherum multiflorum, Beauv., from near Musselburgh, the first record of this grass from Britain.

Epilobium nummularifolium, R. Cunn., in plenty on garden walls at Ardrishaig, Argyllshire : an escape.

Juncus tenuis, Willd., west of West Linton, Peeblesshire.

JAMES M'ANDREW.

Potamogeton prælongus, Wulf., in Orkney. In September 1911 Mr. M. Spence of Durness sent me a specimen of the above species from a loch where it grew with *P. perfoliatus*, L., and *P. filiformis*, Nolte.

It occurs in Shetland!, Caithness!, and Sutherland!. It is also a Faroen species, but is not included in Stefansson's "Flora Islands," 1901.

It is not recorded from Greenland, but occurs at 70° N. lat. at the mouth of the Yenisei (Scheutz., 1888), and nearly as far north in Norway, Alten, 69° 56' (Norman). ARTHUR BENNETT.

Juncus alpinus, Vill., in Kirkcudbright. Mr. G. West records [1] this species from the drier parts of some of the sandy bays of Loch Grennoch by Cairnsmore of Fleet.

It is there associated with "a dwarf form of *Scirpus palustris* about 4 inches high, with short, stout, very scaly rhizomes and few flowering stems."

This loch "is a fine sheet of water 2 miles long by $\frac{2}{3}$ mile wide, at an elevation of 690 feet above sea-level."

This makes eleven Scottish vice-counties in which this species is recorded, and it will eventually, I think, be found in others, especially in Orkney and the Outer Hebrides.

I quite expect to hear of its occurrence in England on the north-west coast, where *Sagina nodosa*, L., var. *moniliformis*, Meyer,[2] was detected this year by Dr. Graebner, and specimens of which Mr. Travis has kindly sent me. This variety occurs on the Friesien Islands, and Bornholm in Denmark, where it grows associated with *Juncus alpinus*.

The *Juncus* is on record for Glamorgan (Dr. Vachell sp.) only, outside Scotland, in the British Isles.

Dr. Buchenau combines under *alpinus*: *J. alpestris*, Hartm.; *J. atratus*, Fries; *J. fusco-ater*, Schreb.; *J. ustulatus*, Hoppe; and *J. nodulosus*, Whlbg. ARTHUR BENNETT.

---

[1] "Cont. Comp. Study of Dominant Phanerogamic, etc., Flora of Aquatic Habit in Scottish Lakes," in Proc. Roy. Soc. of Edin., xxx. (1910), p. 113.

[2] "Hann. Magazine," 1824, p. 169.

Fissidens incurvus Starke, from Kyleakin, Skye. — At the January meeting of the Botanical Society a specimen of this moss in most beautiful fruit was exhibited, sent by Mr. Gerald S. Hoole. Although the moss is not a very rare one, it is seldom that it is seen in as fine fruiting condition as this specimen.

———

Leucobryum pumilum (Michx.) in Britain. I am glad to be able to record that this moss has at last been found in Britain, near Gairloch, Ross-shire. A more detailed notice will appear .in the next number of the Review.   JAMES STIRTON.

———

Notes from Current Literature.

"Botanical Divisions of the British Isles" ("Journal of Botany," November 1911, p. 338).—In this Dr. Moss gives the outlines of the new divisions which it has been thought fit to adopt for the new "Cambridge British Flora." It will be found, amongst other changes, that Clackmannanshire is now merged with Stirlingshire—not with part of Perthshire, as was the case in the Watsonian arrangement. Kinross-shire merges with Fifeshire; and Nairnshire with Elginshire, not with Inverness-shire. The numbers also of the Irish divisions are made to follow consecutively on the British numbers, thus departing from Mr. Praeger's method.

We must confess that as the new divisions, to quote Dr. Moss, "follow, in nearly all cases, ordinary British counties," we fail to see that any good purpose is served by the change. Until some true natural arrangement can be arrived at, it is only causing confusion, and making the consultations of old records more difficult by adding another artificial or parochial scheme to those already in use.

Mimulus moschatus, L., in Aberdeenshire ("Journal of Botany," December 1911, p. 370).—Mr. Wm. Wilson records this from the Wood of Houghton, Alford.

"Some New Forms of Hieracia," by Rev. E. F. Linton, M.A. ("Journal of Botany," December 1911, p. 353).—Several new species and varieties described.

"Mycological Notes," by W. B. Grove, M.A. ("Journal of Botany," December 1911, p. 367), gives notes on *Uromyces flectens*, Lagerheim; *U. Loti*, Blytt; *U. ambiguus*, Lev.; *U. Lilii* (Link.), Fckl., and *Dothidella Betulæ-nanæ* (Karst).

Reviews, Book Notices, etc.

THE TYPES OF BRITISH VEGETATION. By MEMBERS OF THE CENTRAL COMMITTEE FOR THE SURVEY AND STUDY OF BRITISH VEGETATION. Edited by A. G. TANSLEY, M.A., F.L.S. Cambridge University Press. 8vo, cloth, pp. xx + 416, with 36 plates (including more than 60 photographs); 21 figures in the text. 6s. net.

THE authors of this work, and British field botanists in general, are to be equally congratulated on the successful publication of the first handbook, on the distribution of the plant associations of Britain. The book not only forms a condensed record of the pioneer ecological work done in this country, but, from its very attractive nature, is sure to draw recruits to the ranks of plant ecologists. The material has been dealt with in a thoroughly scientific way and from a standpoint which has only been hinted at in a few of the more modern county floras.

Between the title-page and the editor's preface we find a fly-leaf with the following inscription: "To Professor Eugenius Warming, the Father of Modern Plant Ecology, and to Professor Charles Flahault, who, through his pupil Robert Smith, inspired the Botanical Survey of this country, this first attempt at a scientific description of British Vegetation is dedicated, in all gratitude and admiration, by the authors."

On the back of the fly-leaf is a list of the contributors to the book. The first twenty pages include the preface and tables of contents, the remaining four hundred and sixteen pages forming the text, a page of bibliography, and a good index. There are thirty-six excellent plates from photographs and twenty-one figures in the text.

An introductory part and the two following chapters deal shortly with the units of vegetation, their nomenclature, relationships, and classification; the physical characters and climate of the British Isles; and the soils of Scotland, Ireland, England, and Wales. Three small-scale sketch-maps are inserted which bring out clearly the close relations between the distribution of the harder rock-masses, the higher ground, and the areas of greater rainfall; but this close relation between geological denudation, physiography, and climate does not seem to have appealed to the authors of some of the chapters to the degree that might have been expected. The plant formations of the British Isles are stated to be "mainly determined by edaphic factors, *i.e.* by the soil"; but too great importance seems to have been attached to the nature of the geological formations as a direct agent in natural soil production and too little to its influence on geological denudation and physiography, and thus on climate and drainage, and indirectly on the soil through the diverse workings of mechanical disintegration, chemical decomposition, leaching, and the movement of surface and underground waters.

4

The following six chapters are mainly taken up with descriptions of the plant formations of clays and loams, of sandy soils, of the older siliceous soils, and of calcareous soils, where the chief plant associations are believed to have been woodlands. These have been largely destroyed by the hands of man, but interesting descriptions of the former vegetation have been reconstructed from the patches of existing woodland, pastures, and so-called waste ground. The authors hold the view that the chief associations of the plant formations of each of these soil types is a type of woodland having distinctive characters, and that each type of woodland passes into retrogressive associations of scrub and grassland. Thus the plant formation of clays and loams has a chief association of "damp oak-wood," with *Quercus pedunculata* dominant, and retrogressive associations of scrub and "neutral" grassland. That of the sandy soils has a chief association of "dry oak-wood," with *Quercus pedunculata* and *Q. sessiflora* co-dominant, and retrogressive associations of scrub and grass-heath. That of the older siliceous soils, framed to include the soils of the metamorphic and palæozoic non-calcareous rocks, has a chief association of woods, with *Quercus sessiflora* dominant, and retrogressive associations of scrub and "siliceous" grassland. The plant formation of calcareous soils is subdivided into two subformations of the older limestones and of the chalk. The former has a chief association of ash-woods, with subordinate associations of scrub and "limestone" grassland; while the chalk is shown as having associations of beech-woods, ash-woods, yew, scrub, and chalk grassland. The description of the chalk is probably the most interesting, from the naturalist's point of view, that has yet been published.

The heath formation, which seems to the writer to have greater claims as a natural unit of vegetation than any of the above, is described by itself in Chapter IV. It is pictured as having several woodland, heath, and grass associations, and a diagram is given to show the probable genetic relations of the plant communities of the formation of sandy soils and of the heath formation.

In Chapter X. we find an interesting account of the river valleys of East Norfolk, including the physiographical relations of the streams and broads, with their aquatic and fen formations. The conditions affecting the aquatic associations, such as the alkalinity, circulation, aeration, and depth of water, and the factors of light and shelter, are shortly discussed. The various forms of fen and fen-car are considered in detail, and their relations shown by sketch-maps and transects. The great moor formation is described in Chapters XI. and XII. There is an excellent account of the development of the lowland moors of North Lancashire from fens which formerly occupied the position of estuaries and lakes. Marsh and fen are followed by woodland, and this in turn by moorland associations. This description is accompanied by vertical sections of the peat, showing the succession of the vegetation.

The upland moors of the Pennines and the grass moors of the

southern uplands of Scotland are embraced in Chapter XII. In the former is included a full description of the various moorland associations and their distribution, with a short discussion on moorland retrogression, the zonation of associations, and the effects of firing the heather. A brief account of the succession of plant remains in the peat mosses of the Northern Pennines is included, but a wider and more general account might have been expected in view of the amount of work that has been done by Dr. Lewis in this direction.

Chapter XIII. is devoted to the arctic-alpine vegetation. Following a general account of the distinguishing features of this type of vegetation and its distribution is an ecological analysis of the vegetation of Ben Lawers. This is divided into three zones, of which the uppermost forms the zone of arctic-alpine vegetation. Three plant formations are recognised: (1) Arctic-alpine grassland; (2) the formation of the mountain-top débris; (3) the arctic-alpine chomophyte formation. The influence of physiography on the water supply, snowfall, insolation, and shelter from winds, and their effects on the vegetation, are discussed at some length.

In Chapter XIV. the vegetation of the sea-coast is described, and illustrated by beautiful photos. With a general account of the plant associations and conditions of life of the salt-marsh and sand-dune formations, and of the shingle-beach communities, are special descriptions of the salt marshes of the Hampshire coast, the sand-dune vegetation of the Lancashire coast, and the maritime formations of Blakeney Harbour. The relations of the topography and plant associations are shown in interesting sketch-maps.

THE BOTANICAL EXCHANGE CLUB AND SOCIETY OF THE BRITISH ISLES. (*Balance Sheet; Secretary's Report for* 1910.) Report for 1910 by the Editor and Distributor, C. E. Moss, B.A., D.Sc., F.R.G.S., The Botany School, Cambridge. Vol. ii. Published by James Parker & Son, 27 Broad Street, Oxford, 1911. Price 5s. [Pp. 489–610.]

IN this Report, as in recent years, there is to be found much more than the title would lead one to expect. Before the actual report upon the specimens sent in to the Club, are some thirty-eight pages of " Plant Notes for 1910, etc." by that most energetic of botanists, Mr. Druce, who is the treasurer and secretary of the Club. Among these " Notes" there is much of value and interest, and which we feel might well have been published in a more universally read periodical. From amongst these we have selected a few. The number preceding each note refers to Mr. Druce's " List."

" 581 *b. Medicago minima,* Desr., var. *mollissima* (Roth.).—Mollissima, foliolis . . . stipulis basi latioribus apice dentatis, pedunculis bifloris, leguminibus cochleatis 5-gyrosis, spinulis longis apice hamatis, ' Roth. Cat.' 3, 74. Put under *M. minima* as var. *longiseta,*

DC., 'Prod.,' ii. 178, by Rouy and Fonc., and defined : Epines des spires médianes sensiblement plus longues que le diamètre de celles-ci ; plante ordinairement velue-blanchâtres, non glanduleuses. Found at Galashiels, 79, abundantly by Miss Ida Haward. This differs from type *minima* by its soft and whitish pubescence, and by its longer and straighter spines. De Candolle's *longiseta* is said to have longer peduncles bearing several flowers. The Galashiels plant would be put by many botanists as a sub-species. G. Claridge Druce.

"883 *b. Geum rivale*, L., *var. *pallidum*, C. A. Meyer.—With pale greenish flowers, growing over a considerable area in East Lothian. S. Anderson, *in lit.*

"958. *Pyrus pinnatifida*, Ehrh., var. *arranensis* (Hedl.).—Glen Sannox, Isle of Arran ; by some botanists considered a distinct species.

"966 *e. Cratægus oxyacantha*, L., var. *cuneata*, Druce, in 'Journ. Bot.,' 272, 1910. Stylus i. Calyx pubescens. Folia cuneata, angusta, ovato-oblonga, a vertice 3 vel 4 brevibus segmentis divisa. Middlesex. Distinguished from type by the narrower cuneate leaves.

"967 *f. Cratægus oxyacantha*, L., var. *quercifolia*, Druce. Styl. i. Calyces et pedunculi densius hirsuti ; foliis pallide virentibus pubescentibus, in textura molliter—papyraceis, rhomboidis ad bases late cuneatis ; foliorum marginibus in 5–7 segmentis inequaliter divisis, segmentis obscure et diverse crenatis ; ramis floriferis contortis. Kirkcudbright. G. Claridge Druce.

"1297. *Rudbeckia laciniata*, L.—Quite naturalised in Forfarshire ; shown to me by R. D. Corstorphine.

"1912 *bis.* Our British plant appears (teste Prof. Hugo Glück) to be *Veronica aquatica*, S. F. Gray, 'Nat. Coll. Br. Pl.,' ii. 306, 1821, distinguished from *V. Anagallis* by its less crowded racemes and its patent or reflexed peduncles."

There is a note by Mrs. Gregory upon her var. *diversa* of *Viola Riviniana*, Reichenb., but we do not reprint it as there is no full description.

Mr. Druce also gives a most valuable summary of the work done by Dr. Hugo Glück and Fr. Meister upon *Utricularia*, which should be referred to by all those interested in that genus. We extract the following, as it may help our readers in obtaining reliable records for the difficult *U. Bremii* at present doubtfully recorded for Scotland from 81 ? Gordon Moss. 85 ? Loch of Spynie and Moss of Instoch :—

"Dr Glück says the characters which distinguish *Bremii* from *minor* are specific, this species bearing much the same relation to *U. minor* as *U. major* does to *U. vulgaris*. The larger flowers of a darker yellow, with a much larger lip which stares at one, are easy marks of distinction, but in the barren state a greater or less amount of development of leaf-segments and a more sparing development of bladders are not sufficient for separation, although such an appearance may be suggestive.

"As to the leaf-characters of the *Utricularæ* one may add that our

British plants may be divided into three groups by the number of the leaf-tips : (1) very numerous *vulgaris* and *major*; (2) 7-15 *intermedia* and *ochroleuca* ; (3) 14-20 *minor* and *Bremii*.

" Since writing this account of the Utricularias, Dr. Gluck . . . emphasises the impossibility of distinguishing sterile conditions of *U. Bremii* and *U. minor* ; hence the British records of the former based on barren plants appear to be conjectural."

In the second part of the Report by Dr. Moss, there are as usual many interesting notes by most of our best-known botanists. The following are a few :—

" *Viola saxatilis*, Schmidt.—Roadside bank, S. of Loch Rannoch, Perthshire (88), July 1910.—A. Wilson and J. A. Wheldon.

" *Sagina subulata*, Presl.—Sandy roadside, Brown Point, Arran ; v.c. 100, n.c.r. July 1910. V. S. Travis.

" *Tilia* [? *platyphyllos*, Scop.].—Natural wood near Llanvair Discoed, Mon.; v.c. 35, 23rd June 1910. W. A. Shoolbred. The specimens are not typical, and are perhaps from shoots of suckers or of adventitious buds low down on the stems ; and neither flowers nor fruit are sent. However, all the material belongs to the small-leaved lime, *T. cordata* (*T. parvifolia*, Ehrh. ; *T. ulmifolia*, Scop.), and not to the large-leaved lime (*T. platyphylla*), nor to the common lime (*T. europæa*). It is not, I think, usually stated in the floras that the leaves of *T. cordata*, which are borne on suckers, etc., are relatively large (sometimes very large) and have relatively short petioles. *T. platyphylla* may always be distinguished by its hairy twigs, and *T. cordata* by its normal leaves having very long petioles. The cymes of *T. cordata* are not pendent like those of the two other limes ; and they flower later than those of *T. europæa* and *T. platyphylla*. Last year, for example, near Cambridge the flowers of *T. cordata* opened on July 16th, those of *T. europæa* on June 28th, on which latter date those of *T. platyphylla* were already fully out ; and the flowers of *T. europæa* and of *T. platyphylla* were over when those of *T. cordata* appeared. It would be interesting to ascertain whether or not this is invariably the case, as some Continental floras give *T. europæa* as a hybrid of *T. cordata* and *T. platyphylla*. *T. cordata* is also later in coming into leaf than its two allies. C. E. Moss.

" *Juncus tenuis*, Willd. [ref. No. 84].—Roadside in Glen Ogle v.c. 88, September 19th, 1911.—M'Taggart Cowan, jun. Yes, our member Mr. P. Ewing has also sent it me from Ayrshire this year, a N.C.R. for 75. G. Claridge Druce.

" *Zostera nana*, Roth.—Aberlady Bay, Haddington, v.c. 82. . . . Yes, apparently new to Haddington, v.c. 82.—C. E. Salmon. Correct.—P. Graebner.

" *Agrostis alba*, L., var.—Sandy shores, Kildonan, Arran, v.c. 100, July 1910.—W. G. Travis. . . . *A. alba*, var. *condensata*, Hackel, med., var. *coarctata* auct. plur. non *A. coarctata*, Erh.—E. Hackel."

Altogether the Report is a most valuable one, and should be very helpful to all those studying the British flora.

A List of British Roses.   By Major A. H. Wolley-Dod.
"Supp. Journal of Botany," Sept., Oct., Nov., Dec. 1911.

It is a matter for regret that Major Wolley-Dod has been compelled, temporarily at least, to lay aside the study of British roses. We have in Britain so few workers in this most difficult genus that we can ill spare, even for a time, one so enthusiastic and in many respects so competent. We hope that his residence abroad may thoroughly restore his health, and that he may then resume the study with, if possible, even keener zest.

In this paper he gives a revised list of British roses, including names received from Professor Dingler and M. Sudre, with remarks of his own upon each species or variety. He also gives a new analytical key to the sections, and reprints the group keys from his former monographs, but modified and altered in accordance with the new views put forward in the present list. An alphabetical index concludes the paper.

An enormous amount of labour, conscientiously and in many respects very ably performed, was required for the preparation of his former treatises, and even the preparation of the present list must have been a laborious task, especially when it was done under the trying conditions of impaired health, and to some extent against time.

In spite of the experience he has acquired of the thorough worthlessness of most of the micro-species of Deséglise and his school—finding, as he did in most cases, that authentic specimens collected by the authors themselves neither agreed with the descriptions nor with one another,—Major Wolley-Dod still clings to them, and thinks that what we want is not fewer but more names. It is hard to see how confusion can be cleared up by adding more confusion.

There are a great many things in the present list which call for comment, if space were available. Some of them, no doubt, would not have appeared if the conditions under which the list was prepared had been more favourable. I shall refer only to two or three which, if left alone, may lead to serious misconception on the part of British botanists.

M. Sudre, who seems to have an unhappy knack, especially when away from his micro-species, of making surprising determinations—to use a mild term,—named some specimens from Cheshire R. Burnati, Chr., and it is plain that Major Wolley-Dod would have agreed to adopt this name; but as Dr. Dingler, who saw one of the specimens, declared that it could not be R. Burnati, it is excluded for the present. Dr. Dingler "did not say why," but no one who had read Dr. Christ's paper in the "Journal of Botany" for 1876 on the roses of the Maritime Alps would be at a loss to know why. All the roses of this district, he tells us, Canineæ, Rubigineæ, Sepiaceæ, bear the unmistakable impress of a dry and burning Southern climate. "One would say that the Mistral, with its parching breath, has contracted the leafy organs, that the hot and prolonged exposure to the sun has developed in the highest degree the glandulosity, and that the dryness, combined with the heat, has reduced the corolla, and has caused to grow out from the epidermis of the whole plant that enormous quantity of prickles, long, shining, slender, in the shape of sickles and sabres, which give to these roses the strange aspect of a thorny bush of the desert." It is enough to say that *R. Burnati* is one of the varieties peculiar to this region. To another species of the same district, *R. Beatricis*, Burnat and Gremli, M. Sudre assigned an English specimen. Curiously, Major Wolley-Dod rejected this determination, but has applied the name to some other specimens from Surrey. It is a pity these had not also been submitted to Dr. Dingler, for then we should not have had *R. Beatricis* figuring as native in four vice-counties of England.

There is one other point which needs clearing up. At the middle of page 37 we find, "*R. caryophyllacea*, Chr. forma (non Bess.). Two specimens from Catsworth, Hunts (Ley.), are considered by Dingler to agree almost exactly with a form of this species, which he believes may be a hybrid with some Rubiginosa form, but to which he has given the name of *R. tomentella*, var. *anonyma* in his herbarium." On the strength of this Major Wolley-Dod, disregarding Dingler's *R. tomentella*, var. *anonyma*, gives *R. caryophyllacea*, Chr. forma (non Bess.), as occurring in three vice-counties. But what does *R. caryophyllacea* forma (non Bess.) mean in this case? We know what Dr. Christ himself thought, viz. that his species was identical with Besser's. We know what Crépin thought, viz. that its chief varieties were merely very glandular forms of *R. glauca* or *R. coriifolia*. His opinion was shared by Keller, but the latter in his synopsis takes a different view, though he still believes, except for one variety, in its close relation to *R. glauca* and *R. coriifolia*. That variety he joins to Bèsser's species. The others he gives under the name of *R. Rhætica*, Gremli = *R. caryophyllacea*, Chr. (non Besser).

This *R. Rhætica* he describes as a "Berg" rose, with the woolly styles and rising sepals of *R. glauca* or *coriifolia*. It is scarcely found outside of the Lower Engadine and the Tyrol. But there is one variety given by Dr. Christ which is seemingly not noticed by Keller. This was founded on specimens sent from near Gründstadt in Rhenish Bavaria, by Dr. Fries, and was called var. *Friesiana*. This variety has been studied by Dr. Dingler in the living state in its native district. As the result of his examination he dissociates it from *R. caryophyllacea*, Chr., and considers it, no doubt correctly, as a peculiar variety of *R. tomentella*, Lem., giving it in the meantime the provisional name of *R. tomentella*, Lem., var. *anonyma*. To this variety he considers the English specimens from Catsworth closely allied. It is highly probable that the Catsworth specimens, and the others associated with them by Major Wolley-Dod, do belong to a variety of *R. tomentella*, Lem., but it is doubtful if they can be joined as one variety to the Bavarian plant, and it is certain that they can have no relationship to the Berg rose of the Unterengadin, the varieties of which formed the basis of *R. caryophyllacea*, Chr.

PLANT LIFE ON LAND, CONSIDERED IN SOME OF ITS BIOLOGICAL ASPECTS, by Professor F. O. Bower, F.R.S., Glasgow, is one of an excellent " shilling series" from the Cambridge Press, which aims at presenting the more recent aspects of botany in a semi-popular form. Much of the matter is familiar to botanical students, but the author's way of utilising the well-known to illustrate what he wants to convey, makes the book pleasant as well as instructive. Algæ, the Bracken Fern, fossil Cycads, fertilisation and pollination of flowers, and seed-dispersal are dealt with so as to suggest that plant-life begun in water becomes modified for a successful existence on land. While no great effort is made to shirk a certain amount of hard fact, the reader is carried along while the argument is developed, and in the first and last chapters the author appears in a lighter vein. It is refreshing to peruse a book of this type, so different from the back-boneless "popular" books so common now.

CLARE ISLAND SURVEY. Part X. : PHANEROGAMIA AND PTERIDO-
PHYTA. By R. LLOYD PRAEGER, B.A. Dublin: Hodges, Figgis & Co., 1911. Price 4s.

THE Clare Island Survey conducted by a number of Irish naturalists, with the aid of specialists, has now begun to issue results in the form of monographs, one of which, " Phanerogamia and Pteridophyta," by R. Lloyd Praeger, now lies before us. Clare Island is one of the larger of the numerous islands lying in the Atlantic west of Ireland, and it has been examined during the past three years from the points of view of botany, zoology, geology, and other branches. Botanically the island is mainly moorland, but the towering cliffs facing the Atlantic furnish stations for a number of plants, including several well-marked arctic-alpine species. The portions of Mr. Praeger's monograph describing the flora and vegetations are of great interest; they go far beyond merely descriptive lists, and there is a carefully worked-out comparison between Clare Island, other islands near it, and the adjoining parts of the mainland. The distribution of the chief types of vegetation is illustrated by a map. Of more general interest is the discussion of the origin of the flora, in which the part played by water, wind, birds, man, and geological factors are all so carefully considered that the memoir is really a work of reference on this—a subject of great importance in connection with similar problems on the Western Isles of Scotland. Is it too much to hope that in the not very distant future a similar survey will be instituted on some of the Scottish islands?

Telegraphic Address: "LARCH," EDINBURGH.   Telephone No. 2034.

Awarded 4 Gold and 4 Silver Medals at recent Flower Shows held in Edinburgh.

DAVID W. THOMSON

Nurseryman and Seedsman

113 GEORGE STREET
—— EDINBURGH. ——

Seed Warehouse—113 GEORGE STREET.

**Selected Vegetable and Flower Seeds,
Bulbs and Forcing Plants,
Retarded Bulbs and Plants,
Garden Tools, Manures, &c.**

Catalogues Post Free on Application.

URSERIES—Granton Rd. and Boswall Rd.

AN EXTENSIVE AND WELL-GROWN STOCK OF

**Forest Trees of all kinds,
Ornamental Trees and Shrubs,
Rhododendrons and Flowering Shrubs,
Game Covert Plants,   Fruit Trees,
Roses and Climbing Plants,
Herbaceous and Alpine Plants,**

ALL IN SPLENDID CONDITION FOR REMOVAL.

INSPECTION INVITED.

Catalogues Post Free on Application.

Contents

Printed by BoD™in Norderstedt, Germany